Decision-Making in Enviro...

Also available from E & FN Spon

Air Pollution
An introduction
J. Colls

Clay's Handbook of Environmental Health
Edited by W.H. Bassett

Dams and Disease
Ecological design and health impacts of large dams, canals and irrigation systems
W. Jobin

Environmental Health Procedures
W.H. Bassett

Determination of Organic Compounds in Natural and Treated Waters
T.R. Crompton

Determination of Organic Compounds in Soils, Sediments and Sludges
T.R. Crompton

International River Water Quality
Pollution and restoration
Edited by G. Best, E. Niemirycz and T. Bogacka

Microbiology and Chemistry for Environmental Scientists and Engineers
J.N. Lester and J.W. Birkett

Monitoring Bathing Waters
A practical guide to the design and implementation of assessments and monitoring programmes
Edited by J. Bartram and G. Rees

The Coliform Index and Waterborne Disease
Problems of microbial drinking water assessment
Edited by C. Gleeson and N. Gray

Toxic Cyanobacteria in Water
A guide to their public health consequences, monitoring and management
Edited by I. Chorus and J. Bartram

Urban Traffic Pollution
Edited by D. Schwela and O. Zali

Urban Drainage
D. Butler and J. Davies

Water and the Environment
Innovative issues in irrigation and drainage
Edited by L.S. Pereira and J. Gowing

Water: Economics, management and demand
Edited by B. Kay, L.E.D. Smith and T. Franks

Water Policy
Allocation and management in practice
Edited by P. Howsam and R.C. Carter

Water Pollution Control
A guide to the use of water quality management principles
R. Helmer and I. Hespanhol

A Water Quality Assessment of the Former Soviet Union
Edited by V. Kimstach, M. Meybeck and E. Baroudy

Water Quality Assessments, 2nd edition
A guide to the use of biota, sediments and water in environmental monitoring
Edited by D. Chapman

Water Quality Monitoring
A practical guide to the design and implementation of freshwater quality studies and monitoring programmes
Edited by J. Bartram and R. Ballance

Water Resources
Health, environment and development
Edited by B. Kay

Water Wells: Monitoring, maintenance, rehabilitation
Proceedings of the International Groundwater Engineering Conference, Cranfield Institute of Technology, UK
Edited by P. Howsam

Decision-Making in Environmental Health

From evidence to action

Edited by
C. Corvalán, D. Briggs, and G. Zielhuis

Published on behalf of the

**WORLD HEALTH
ORGANIZATION**

London and New York

First published 2000 by E & FN Spon
11 New Fetter Lane, London EC4P 4EE

Simultaneously published in the USA and Canada
by E & FN Spon
29 West 35th Street, New York, NY 10001

E & FN Spon is an imprint of the Taylor & Francis Group

Printed and bound in Great Britain by TJ International Ltd, Padstow, Cornwall.

Publisher's Note
This book has been prepared from camera-ready copy provided by the editors.

British Library Cataloguing in Publication Data
A catalogue record for this book is available from the British Library.

Library of Congress Cataloging in Publication Data
Decision-making in environmental health: from evidence to action / edited by C.
Corvalán, D. Briggs, and G. Zielhuis.
        p.cm.
    "Published simultaneously in Canada."
    Includes bibliographical references and index.
    ISBN 0-419-25940-6 (hb : alk.paper) — ISBN 0-419-25950-3 (pbk. : alkl.paper)
    1. Environmental health—Decision making. 2. Environmental health. I. Corvalán,
C.
    (Carlos) II. Briggs, D. (David) III. Zielhuis, G. (Gerhard)
    RA566.D43 2000
    615.9'02—dc21   00-021614

TABLE OF CONTENTS

FOREWORD

This book evolved from the need to address a number of fundamental questions relating to environmental health for which there were no simple answers. These questions ranged in scope and depth, from issues related to basic statistics on health and the environment to the use of information in the management of problems associated with environmental health. Many of these questions were concerned with the way in which information is, or can be, used to help address environmental health problems, and with the role and value of environmental health indicators. Examples of these questions are:

How can one collect and present information which is useful in shaping and making decisions at the local level?

What does a national indicator (e.g. infant mortality rate or access to water and sanitation) mean in the face of large disparities at the sub-national level?

Why is it not always possible to quantify indicators at the sub-national level, if national-level indicators exist?

What do environmental exposure indicators mean beyond the local level, where people are affected?

Such questions indicate a need to address issues relating to the requirements and use of local-level information. Other questions were of a more technical nature, for example:

What is the health impact in terms of morbidity and mortality of a given environmental exposure?

How does the impact vary according to age, gender, geographical location and socio-economic group?

How are environmental health problems ranked and prioritised at the local level?

Further questions referred to policy and decision-making issues, for example:

How does the environmental health decision-making process operate locally?

How are locally collected data transformed into information and used in decision-making, or if such information is not used, what are the reasons?

This book addresses these and other related issues. It proposes a model for decision-making in environmental health based on the involvement of relevant stakeholders and the use of scientifically sound data and appropriate analytical methods. It also proposes a framework for understanding environmental health problems and their effects in a manner that allows inter-disciplinary and intersectoral approaches to action. Finally, the book recommends the development of local environmental health information systems for the collection of locally relevant data, with emphasis on simplification to avoid overloading such systems.

The link between the environment and human health has been suspected for centuries; there is now widespread consensus that healthy environments are prerequisites for human existence and health. However, the link between development activities and their impact on health and the environment is a more recent issue. At the Earth Summit, held in Rio de Janeiro in 1992, it was recognised that both insufficient development, leading to poverty, and inappropriate development, leading to over-consumption, could result in severe environmental health problems. In all countries, information about health and the environment at different levels (e.g. village, city, province or country) is necessary in order to support the management and decision-making process in relation to environmental health. Providing relevant information, in a form that all those involved can understand and accept, within the constraints of time and other resources, is thus a major challenge. It is not simply a matter of collecting data. In order to be useful, environmental health information should be pertinent, and sufficiently accurate and usable by all those involved in decision-making, from the public to political leaders. The decision-making process requires information that is directly relevant to the task in question, the translation of this information into a consistent and coherent form, and the presentation of the information in a manner that is appropriate and acceptable to the different users. This book addresses these issues in detail.

This book will be useful to researchers in public health, epidemiology and the social sciences. It will also be useful to those working in government institutions concerned with environmental health, particularly those responsible for collecting and analysing data as part of local or national information systems.

World Health Organization

ACKNOWLEDGEMENTS

The World Health Organization wishes to express its appreciation to all those whose efforts made this book possible. An international group of authors contributed to this book, and for some chapters more than one author and their collaborators provided material. It is difficult to identify precisely the contribution of each individual author and therefore the principal contributors are listed together below:

Françoise Barten, Nijmegen Institute for International Health, Nijmegen University, Nijmegen, Netherlands (Chapters 2 and 8)

David Briggs, Nene Centre for Research, University College Northampton, Northampton, England (Chapters 1, 3, 4, 5, 7 and 9)

Carlos Corvalán, Department of Protection of the Human Environment, World Health Organization, Geneva, Switzerland (Chapters 1, 2, 3, 5, 8 and 9)

Ken Field, School of Environmental Sciences, University College Northampton, Northampton, England (Chapter 7)

Tord Kjellström, Department of Community Health, University of Auckland, Auckland, New Zealand (Chapters 1 and 3)

Markku Nurminen, Department of Epidemiology and Biostatistics, Finnish Institute of Occupational Health, Helsinki, Finland (Chapters 5 and 6)

Tuula Nurminen, Department of Epidemiology and Biostatistics, Finnish Institute of Occupational Health, Helsinki, Finland (Chapter 5)

Gerhard Zielhuis, Nijmegen University, Nijmegen, Netherlands (Chapters 1, 2, 8 and 9).

Acknowledgements are also due to the following people, who contributed to the planning and implementation of the field studies: Simon Lewin, Medical Research Council of South Africa, South Africa; A. Mukherjee (and colleagues), Centre for Study of Man and the Environment, Calcutta, India; Manuel Salinas (and colleagues), Universidad Catlica, Santiago, Chile; Angel Sanchez (and colleagues), Universidad Nacional Autnoma de Nicaragua, Nicaragua; Andre Soton (and colleagues), Centre Régional pour le Développement et la Santé, Cotonou, Benin; Carolyn Stephens, London School of Hygiene and Tropical Medicine, London, England; Ronald Subida,

University of the Philippines, Manila, Philippines; and Elma Torres, University of the Philippines, Manila, Philippines.

The World Health Organization also thanks the following people, who provided ideas and comments: Marco Akerman, Brazil; Margaret Conomos, Environmental Protection Agency, USA; Greg Goldstein, Department of Protection of the Human Environment, World Health Organization, Geneva, Switzerland; Hiremagalur Gopalan, United Nations Environment Programme, Nairobi, Kenya; Jim Leigh, National Occupational Health and Safety Commission, Australia; Steven Markowitz, City University of New York, NY, USA; Tony McMichael, London School of Hygiene and Tropical Medicine, London, England; Barry Nussbaum, Environmental Protection Agency, USA; Harris Pastides, University of South Carolina, SC, USA; Eugene Schwartz, USA; Jacob Songsore, University of Ghana, Ghana; Judy Stober, Intergovernmental Forum on Chemical Safety, World Health Organization, Geneva, Switzerland; and John Wills, University College Northampton, Northampton, England.

Thanks are also due to Deborah Chapman and her assistants for editorial assistance, layout and production management. As the editor of the WHO co-sponsored series of books dealing with various aspects of environmental health management, Deborah Chapman was also responsible for ensuring compatibility with other books in the series.

Special thanks are due to the Environmental Protection Agency (USA) and the United Nations Environment Programme, which provided financial support for the activities related to the production of this book.

Chapter 1[*]

HEALTH AND ENVIRONMENT ANALYSIS

1.1 Background

Human exposure to pollutants in the air, water, soil and food — whether in the form of short-term, high-level, or long-term, low-level exposure — is a major contributor to increased morbidity and mortality. However, the disease burden attributable to these exposures is not known with any degree of certainty because levels of general environmental pollution fluctuate greatly, methods for analysing the relationships are incompletely developed, and the quality of available data is generally poor. Precise measures of the association between pollution levels and health outcomes are therefore rare. Exposure to environmental pollution is also usually involuntary. People may be unaware of this and/or its possible effects; as a result they may exert little control over their risk of exposure. Biological and chemical agents in the environment are nevertheless responsible for the premature death or disablement of millions of people worldwide every year (WHO, 1992). It has recently been estimated that almost one quarter of the global burden of disease is attributable to environmental factors (WHO, 1997). This estimate, which is based on published data (Murray and Lopez, 1996), was made by attributing an environmental causal fraction to each disease category with a known environmental link. The ability to link health and environmental data, and thereby to determine the relationship between levels of exposure and health effects, is clearly vital to control exposure and protect health. Decision-makers need information on the health effects attributable to environmental pollution in order to assess the implications of their decisions, to compare the potential effects of different decisions and choices, and to develop effective prevention strategies.

Standards and guidelines against which to assess levels of environmental pollution are now widely available. For example, the World Health Organization (WHO) has developed environmental quality guidelines for various pollutants in the air (WHO, 1987), drinking-water (WHO, 1993),

---

[*] *This chapter was prepared by C. Corvalán, T. Kjellström, G. Zielhuis and D. Briggs*

food (FAO/WHO, 1989) and workplace (e.g. WHO, 1980, 1986). These guidelines are based on epidemiological and toxicological studies and indicate the maximum environmental levels, or the maximum levels of human exposure, considered acceptable in order to protect human health. Nevertheless, individual susceptibility to pollution varies, to the extent that it is possible that some individuals may experience adverse effects at levels below the maximum recommended levels. Moreover, in many areas of the world these levels are frequently exceeded, in some places by as much as several times the recommended levels, and reduction of human exposure may be difficult or very costly. Adverse effects on human health are therefore likely to continue to be observed in these areas. In such situations, analysis of data on human health and the environment provides a valuable tool for obtaining estimates of the health impact of pollution, which can be used to set priorities for action.

Many epidemiological studies have been undertaken to analyse the relationship between specific forms of environmental pollution and health effects. Most of these studies have been conducted in developed countries, and the methods used may not be applicable to other settings, especially if high quality data are unavailable or too expensive to collect. Major problems often exist in obtaining data on health and particularly on environmental exposure at the individual level. As a consequence, it is normally necessary to rely on so-called "ecological" methods, in which the statistical unit of observation is a population rather than an individual (Rothman, 1986; Beaglehole et al., 1993; see Chapter 6).

A serious limitation in conducting epidemiological studies concerns the measurement of exposure in individuals. Routinely collected environmental data are widely available in most countries and, where relevant, can be used as a proxy for exposure data. For example, monitoring networks provide data on pollution levels at specific sites, which can be used to characterise average exposures for geographical regions. Environmental data are also often compared with guideline values or maximum recommended levels in order to determine levels of compliance with prevailing policies. The data are, however, rarely used to quantify the potential health effects. Equally, although many countries routinely collect data on health outcomes in the form of morbidity and mortality statistics, attempts are rarely made to link the data to environmental or other factors in order to attribute outcomes to their cause.

1.2   Tools for analysis and interpretation

Linking environmental and health data offers considerable benefits, but also poses many dangers if not carefully carried out. In linking such data it is all too easy to overlook the statistical problems and inconsistencies of the different data sets, or to misinterpret their apparent relationships. Valid linkage thus relies on the use of both valid data and appropriate linkage methods.

Numerous methods for data linkage have been developed in many different areas of application. Their suitability for linking environmental and health data, however, is often limited and always needs to be assessed carefully. Two important criteria must be considered in this context. First, the methods must be politically acceptable. This means that they must be simple, inexpensive to implement, and operable with the available data, thus allowing rapid assessment. If the methods are overly complex, requiring extensive resources and collection of large amounts of additional data, few developing countries will be able to apply them, and even in developed countries their use may be costly and result in delays in action. Second, if the results are to be accepted as a basis for action, the methods must be scientifically credible and statistically valid. This means that they should be accurate, sensitive to variations in the data of interest and unbiased. Simple, crude methods should produce results that agree with those obtained from more detailed studies, for which the statistical precision can be quantified.

In practice, these requirements are rarely met in full. If they were, there would hardly be a need for individual-level studies. Nevertheless, simple methods may still have considerable value. Results from ecological studies, for example, are useful if the potential biases can be identified, evaluated and shown to be small. At the very least, the results should help to identify areas or issues requiring more detailed investigation. Countries where detailed, individual-level studies have not been performed also urgently need access to methods which can help to shed light on the extent and health effects of specific forms of environmental pollution. Priority should be given to the development of research capabilities in developing countries for this purpose (Environmental Research, 1993).

Where detailed information on the exposure–response relationship of specific pollutants is available, Quantitative Risk Assessment (QRA) techniques, based on epidemiological data, can be used to estimate the impact of exposure on different populations without the need for new substantive research (for further information, see Romieu et al., 1990; Nurminen et al., 1992; Ostro, 1996). This implies knowledge about exposure, the population at risk and the health effects associated with exposure in the form of a dose–response function derived from epidemiological studies (i.e. pooled study results) (Goldsmith, 1988; Smith, 1988; Hertz-Picciotto, 1995; Smith and Wright, 1995; Wartenberg and Simon, 1995). Because of limitations in available research data, QRA can often be applied only by extrapolating study results from one country (usually developed) to other countries (usually less developed). The fact that the range of exposure levels and the distribution of other conditions likely to affect health outcomes may differ substantially between populations inevitably limits the validity of this approach. In addition, assessments can only be carried out reliably for pollutants for which well researched exposure–response relationships have been established. Even then, uncertainty regarding the assumed association between environmental pollution levels and the actual exposures in individuals is a major constraint.

QRA remains the only tool available for estimating the health outcomes of environmental pollution in areas where health monitoring is not undertaken, or where the quality of the data collected is poor. It is also the only feasible approach for obtaining crude estimates of health impacts in very large population groups. The development and application of well tested methods of risk assessment is therefore an important priority. It is equally important to describe the risks of exposure which exist to decision-makers and the community in a meaningful way (Rose, 1991).

1.3   Health and Environment Analysis for Decision-Making Project
The Health and Environment Analysis for Decision-Making Project (HEADLAMP) (Corvalán and Kjellström, 1995) is aimed at addressing some of the limitations outlined above in the information currently available to support environmental health policies. Its overall purpose is to make valid and useful information on the local and national health impact of environmental hazards available to decision-makers, environmental health professionals and the community, in order to promote effective action to prevent or reduce environmental health problems. To this end, it is designed to indicate environmental health trends, as a basis for defining appropriate policies and for assessing the value and performance of these policies over time. It also aims to encourage local and national capacity-building, as a means of enabling environmental health issues to be tackled more effectively at the appropriate level.

HEADLAMP takes a deliberately interdisciplinary and intersectoral approach. It uses a combination of methods from environmental epidemiology (including human exposure assessment) and other health and environmental sciences to produce and analyse data, to convert these data into information, and to present this information so that it can be understood and acted upon by those responsible for environmental health protection. Three principles define the HEADLAMP process:

1.   HEADLAMP is based on scientifically established relationships between environmental exposure and health effects. This approach has proved successful in surveillance systems for the prevention and

control of occupational diseases, and has been shown to be most effective when based on a sound set of data relating to both exposure and health outcomes (Thacker *et al.*, 1996).

2. HEADLAMP makes use of environmental health indicators to assess and monitor the environmental health situation, to help define the actions which need to be taken, and to inform those concerned. The indicators are chosen according to the issue requiring investigation, which in turn determines the data and method needed. The development of appropriate environmental health indicators is clearly integral to the HEADLAMP approach.

3. As far as possible, HEADLAMP uses routinely collected data. A major advantage of this approach is its cost-effectiveness. Data collection is expensive, and it is therefore important to obtain the maximum value from data through their repeated and effective use. To measure the relevant environmental health indicators, it may also be necessary to collect additional data. In these situations HEADLAMP encourages the use of appropriate, low-cost techniques.

The key to HEADLAMP is clearly information. Attempts to use information to support health intervention and policy are not new. Current health information systems, however, have been criticised because of the extra demands they impose on health workers, their tendency to centralise information (often in ways which make it inaccessible to many potential users), the failure to analyse adequately the collected data for use in planning, the aggregation of data which masks areas where action is required, and the failure to build links with other sectors (de Kadt, 1989). HEADLAMP is designed to avoid these weaknesses and limitations. It brings together not only the different sectors but also the many different stakeholders involved, including the community and local decision-makers. It builds upon existing health and environmental information systems and promotes the use of existing data, thereby allowing a feedback process to data collectors regarding its quality and the need for additional data. It also encourages data to be translated into information which can be used by different stakeholders and can act as an aid to decision-making. Moreover, HEADLAMP operates at the local level, avoiding problems of information centralisation and aggregation at higher levels. Through the implementation of the Programme of Action for Sustainable Development (Local Agenda 21) (United Nations, 1993), local governments are likely to take the lead role on environmental health at the local level (Williamson, 1996). HEADLAMP is thus a potentially useful tool for action at this level.

1.4   The HEADLAMP process

HEADLAMP has been developed as a practical methodology to address the adverse effects of specific environmental conditions on human health at the local level. Application of the HEADLAMP process follows three stages, as follows (see Figure 1.1):

1.   *Definition of the problem.* The issue(s) to be addressed may be defined initially in many different ways: for example, through the concerns of the local community, as a result of local investigations, or as a consequence of priorities set at a wider level (e.g. as a local response to a National Environmental Health Action Plan). In each case, however, an essential prerequisite is a set of known links (validated by previous research) between a defined environmental factor and its associated health outcomes. Basic information needed to address this issue is identified at this stage. The participation of all relevant stakeholders concerned is also necessary, because the process is intersectoral, and aims to draw together not only the health agencies but also other sectors related to the problems at hand. Together, these various stakeholders can help to redefine the issue in clearer terms and to provide practical guidance and help in developing an appropriate methodology and locating relevant data.

2.   *Compilation, assessment and quantification of relevant environmental health indicators.* During this stage, detailed data requirements are specified, taking account of the specific setting in which the analysis is being conducted, and the limitations of data availability. These data are obtained as far as possible from available routine data sources, but may be supplemented where necessary through the implementation of purposely designed, rapid surveys. Once collected, these data are then processed and analysed to provide information on the environmental health issues of concern. The variables produced through this process comprise the environmental health indicators. Depending on the problem and/or feasibility of obtaining all the relevant data, environmental health indicators may be derived either from health data (e.g. specific morbidity rates attributable to definable environmental factors) or environmental data (e.g. pollution levels with known human health implications). Where appropriate, these indicators are then linked (usually at an aggregate level) to provide further information on the environmental health situation.

3.   *Formulation and implementation of appropriate policies.* At this stage, the trends and patterns shown by the environmental health indicators are interpreted and, based on this interpretation, appropriate

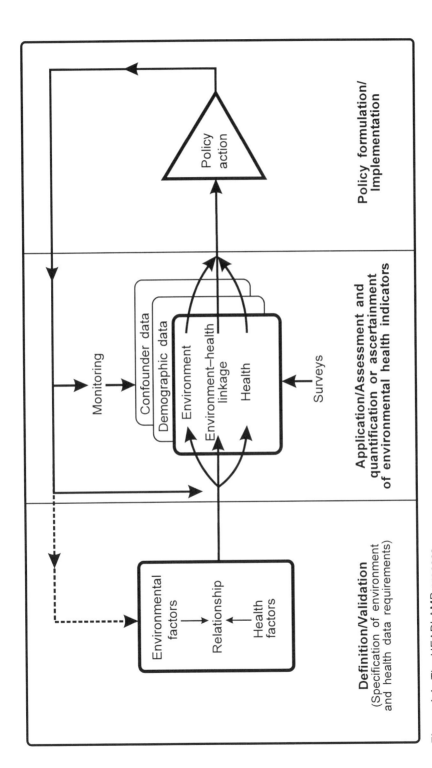

Figure 1.1 The HEADLAMP process

policy responses are defined, the relevant stakeholders and actors are informed, and the actions implemented.

In this context, the HEADLAMP process needs to be seen, not as a one-off activity, but as part of a continuing cycle of monitoring and policy review in which repeated assessments of health and environmental status are used first to develop, and then to revise, effective actions to reduce exposures likely to have adverse health effects. Thus, repeated assessments may be undertaken at appropriate intervals in order to monitor changes in health and/or environmental status and to detect any trends or patterns which may exist. These assessments also allow the effects of policy implementation to be monitored and can help to define any changes which might be needed. They also provide a source of environmental health information for the public and other stakeholders. Where appropriate, a decision to cease monitoring activities might also be taken once pre-set targets have been met on a sustained basis.

1.5   Summary

Application of the HEADLAMP approach is aimed at improving protection against environmentally related disease and the promotion of a healthy environment. Reduction of exposure requires investment by people and authorities. Given the shortage of resources for essential development activities in virtually all countries, scientifically sound and convincing information is essential to motivate and justify such investment. The information required is likely to include clear specification of the problem, its importance, and the costs and benefits of possible responses. Providing this information requires the availability of suitable methods of data analysis and linkage, as well as of indicators which can express the results of these analyses in terms which are understandable and relevant to decision-makers. Methods of data linkage and use of environmental health indicators can, therefore, be invaluable tools for policy-making and management.

The implementation of HEADLAMP activities at the local level should complement and support existing environmental health efforts. If effective decision-making and actions can be sustained and multiplied in many local situations, a significant impact at the national and global levels is expected.

1.6   References

Beaglehole, R., Bonita, R. and Kjellström, T. 1993 *Basic Epidemiology*. World Health Organization, Geneva.

Corvalán, C. and Kjellström, T. 1995 Health and environment analysis for decision-making. *World Health Statistics Quarterly*, **48**, 71–7.

de Kadt, M. 1989 Making health policy management intersectoral: issues of information analysis and use in less developed countries. *Social Science Medicine*, **29**, 503–14.

Environmental Research 1993 Environmental health agenda for the 1990s. Summary of workshop, October 16, 1991, Santa Fe, New Mexico. *Environmental Research*, **63**, 1–15.

FAO/WHO 1989 *Summary of Acceptances: Worldwide and Regional Codex Standards. Codex Alimentarius Part 1, Rev. 4.* Food and Agriculture Organization of the United Nations, Rome/World Health Organization, Geneva.

Goldsmith, J.R. 1988 Commentary on Epidemiologic input to environmental risk. *Archives of Environmental Health*, **43**, 128–9.

Hertz-Picciotto, I. 1995 Epidemiology and quantitative risk assessment: a bridge from science to policy. *American Journal of Public Health*, **85**, 484–90.

Murray, C.J.L. and Lopez, A.D. [Eds] 1996 *The Global Burden of Disease: a Comprehensive Assessment of Mortality and Disability from Diseases, Injuries and Risk Factors in 1990 and Projected to 2020.* Harvard University Press, Cambridge, Massachusetts.

Nurminen, M., Corvalán, C., Leigh, L., Baker, G. 1992 Prediction of silicosis and lung cancer in the Australian labour force exposed to silica. *Scandinavian Journal of Work, Environment and Health*, **18**, 393–9.

Ostro, B. 1996 *A Methodology for Estimating Air Pollution Health Effects.* World Health Organization, Geneva.

Romieu, I., Weitzenfeld, H. and Finkelman, J. 1990 Urban air pollution in Latin America and the Caribbean: health perspectives. *World Health Statistics Quarterly*, **43**, 153–67.

Rose, G. 1991 Environmental health: problems and prospects. *Journal of the Royal College of Physicians of London*, **25**, 48–52.

Rothman, K.J. 1986 *Modern Epidemiology.* Little Brown & Co., Boston.

Smith, A.H. 1988 Epidemiologic input to environmental risk assessment. *Archives of Environmental Health*, **43**, 124–7.

Smith, A.H. and Wright, C. 1995 Environmental risk assessment using epidemiologic data. In: *Proceedings of the Conference and Workshop Host Factors in Environmental Epidemiology.* Cracow, Poland, June 11–14, 9–25.

Thacker, S., Stroup, D., Parrish, R. and Anderson, H. 1996 Surveillance in environmental public health: issues, systems and sources. *American Journal of Public Health*, **86**, 633–41.

United Nations 1993 *Agenda 21: Programme of Action for Sustainable Development.* United Nations, New York.

Wartenberg, D. and Simon, R. 1995 Comment: integrating epidemiologic data into risk assessment. *American Journal of Public Health*, **85**, 491–3.

WHO 1980 *Recommended Health-Based Limits in Occupational Exposure to Heavy Metals.* WHO Technical Report Series 647, World Health Organization, Geneva.

WHO 1986 *Recommended Health-Based Limits in Occupational Exposure to Selected Mineral Dusts*. WHO Technical Report Series 734, World Health Organization, Geneva.

WHO 1987 *Air Quality Guidelines for Europe*. WHO Regional publications, European Series No. 23, World Health Organization, Copenhagen.

WHO 1992 *Our Planet Our Health*. Report of the WHO Commission on Health and Environment. World Health Organization, Geneva.

WHO 1993 *Guidelines for Drinking-Water Quality. Volume 1. Recommendations*. World Health Organization, Geneva.

WHO 1997 *Health and Environment in Sustainable Development Five Years After the Earth Summit*. World Health Organization, Geneva.

Williamson, J.D. 1996 Promoting environmental health. *British Medical Journal*, **312**, 864.

Chapter 2[*]

REQUIREMENTS FOR SUCCESSFUL ENVIRONMENTAL HEALTH DECISION-MAKING

2.1 The essence of environmental health decision-making

Environmental health programmes aim at preventing needless morbidity and mortality by protecting people from unnecessary exposure to environmental hazards. Unfortunately, despite the increasing knowledge about potentially harmful exposures, preventative action is often slow to follow. The mismatch between knowledge and application or translation is often most acute in developing countries, where environmental and occupational exposures often exceed national and international guideline levels, yet where corrective action to control these problems is limited. To reduce this growing deficit of action, research findings and monitoring data need to be translated more effectively and efficiently into public health practice. This requires the provision of the right type of information, and its communication to decision-makers in an easily understandable and appropriate form. Better tools to help decision-makers use the available epidemiological data also need to be developed. It has been argued that decisions are hardly ever taken because of evidence, but instead that evidence is usually used to support existing positions and policies (Hunt, 1993). Under this paradigm, individual decision-makers have been able to dictate actions on the basis of what is seen as politically favourable rather than responding to society's concern. Increasingly, however, ideals such as equity in health, environmentally sustainable development, public accountability and liability, and the formation of partnerships and involvement of the community and other important groups are changing this paradigm.

Decision-making is, certainly, a complex process. It involves choosing among alternative ways of meeting objectives. Implicit in this definition is the notion that there are a number of alternatives, and that their effects can be measured or estimated and compared (Warner *et al.*, 1984). This, in turn, implies that there is adequate information on which to make an informed choice. Often, however, these ideals are not met. Commonly, there is limited

* *This chapter was prepared by C. Corvalán, F. Barten and G. Zielhuis*

or inadequate information on the potential impact or costs of various policy alternatives, or even on what policy options are available. There may be confusion between the risks and benefits of specific interventions; for example reducing water chlorination to reduce the risk of cancer may increase the risk of waterborne diseases (Graham and Wiener, 1995). Those who gain and those who lose from the various actions may also differ, so that social values and scales have to be introduced to allow the options to be traded off against each other, and a decision reached. Together, this uncertainty and conflict may produce diverse conclusions about the "best alternative" when viewed by different observers.

The amount and type of information available is a major driving force for policy. The importance of information for decision-making is discussed in more detail in the following sections.

2.2   Historical development of environmental health decision-making

The links between the environment and public health have been known or at least suspected for many centuries. During the reign of Edward I of England (1272–1307), for example, it was recognised that the burning of "sea coal" produced "*so powerful and unbearable a stench that, as it spreads throughout the neighbourhood, the air is polluted over a wide area*" and this was found to be "*to the detriment of their [the citizen's] bodily health*" and therefore forbidden by direct order of the King (Wilson and Spengler, 1996). History has also shown that not all decisions taken are rapidly implemented. In this case the problem did not end with the signing of the King's orders. In the following reign, Edward II (1307–27) ordered air polluters to be tortured; half a century later, Richard II (1377–99) opted for the restriction of coal use through taxation (Wilson and Spengler, 1996).

This example is one case showing that concentration of efforts on the causes of ill health (health determinants) rather than on the health effects makes good sense. It also shows that the different decisions do not always lead to successful implementation of preventative measures. In many other situations the benefits of focusing on causes rather than effects has been less clear, in part because the links between some determinants and public health are not always direct. The effect of poverty on health status provides a classic example (WHO, 1996). In fact, the first systematic and convincing assessment of the efficacy of determinant-based interventions was probably that by McKeown (1976). This not only questioned the role of medicine in the improvement of health, but also presented evidence that the decline of mortality and morbidity in the past century was due primarily to limitation of family size, improvement of nutrition, a healthier physical environment (e.g. hygiene) and specific preventative measures, rather than a result of therapeutic action. From these observations McKeown (1976) and others infer that successful public health interventions are those which concentrate efforts on improving human environments, both physical and social, and claim that this is best achieved through the combined efforts of society at large, and not by the health sector on its own (Brown *et al.*, 1992). Although not without challenge (Sundin, 1990), the analyses of McKeown (1976) stimulated a revived interest in public health and preventative medicine, and a shift away from the therapeutic view which tended to dominate health policy in previous decades (Ashton, 1992).

The shift in focus towards an environmental perspective of health was echoed and endorsed by government reports and global health policy. In Canada in the mid-seventies, for example, an important report on the health of the population, known as the *Lalonde Report*, argued that future improvements in health status would be due mainly to improvements in the environment, lifestyles, and the increasing knowledge of human biology (Lalonde, 1974). This approach to health policy also allowed the active involvement of other disciplines and sectors in the health arena (O'Neill, 1993). In 1978, a similar change in thinking at the global policy level was witnessed. In that year, the first International Conference on Primary Health Care held in Alma Ata (former USSR) launched a major public health movement, known as "Health for All", which emphasised equity in health, health promotion and protection, intersectoral action, community participation and primary health care (WHO, 1978). The "Health for All Strategy" has been a major force for global action on health since then.

The links between development, environment and public health have taken global prominence over the past decade, particularly since the emergence of "sustainable development" as a guiding principle for policy, and the adoption in 1992 of Agenda 21 (United Nations, 1993) at the United Nations Conference on Environment and Development (UNCED). This has also helped to focus policy attention on environmental health determinants, particularly with respect to the impact of pollution and resource depletion.

The interactions between development, environment and health have been discussed in different contexts (e.g. Bradley, 1994; Warford, 1995). The links between these different areas is both varied and complex. In the context of tropical development, for example, Bradley (1994) cites twelve possible interactions where the activities of one area may favour or impede the functioning of each of the other two. So-called "win-win" situations would occur when both actions aimed at improving the environment or development also favour health. Examples are the improvement of water quality, in the first case, and reduction of poverty in the second. In turn, initiatives to improve health may favour both the environment and development.

Sustainable development has been defined as "*development that meets the needs of the present without compromising the ability of future generations to meet their own needs*" (World Commission on Environment and Development, 1987). Developments which jeopardise human health, whether through pollution or resource depletion, are clearly not sustainable. Principle 1 of the Rio Declaration, for example, clearly stated the case by placing human beings "*at the centre of concerns for sustainable development*" (United Nations, 1993). Chapter 6 of Agenda 21 takes this principle further by emphasising the fundamental commitment within sustainable development of "*protecting and promoting human health*".

It is widely accepted that until now, science and technology have been able to compensate for the world's unsustainable practices. Improvements in prospecting and production and the development of substitutes have generally masked the loss of environmental resources which has been taking place. Reliance on scientific research and technological improvements, however, disregards the risk that human pressures will ultimately outgrow the rate of "response" that science and technology can provide (McMichael, 1993). Adherence to the principles of sustainable development implies that tomorrow's science can no longer be relied upon to solve problems created today. Sustainable development implies both environmental and human health protection.

Viewed in these terms, it is clear that sustainable development is a narrow and fragile entity. If resources are not used efficiently and effectively, development may suffer and many in the world will be forced to remain at an unnecessarily low standard of living. Health, if not the environment, will certainly be impaired. On the other hand, even a slight excess rate of resource use, if continued for long periods, will deplete the world's resources and damage the environment, again to the detriment of human health. Sustainable development thus requires delicate guiding of human action, and well-targeted and well-informed policy. Information is therefore essential to agree the goals, to guide actions, and to assess progress in the desired direction.

2.3   Examples of successful environmental health decision-making

Taking decisions in general, and decision-making on environmental health in particular, is a complex process, involving people at all levels of society. This can be illustrated by the following example.

A government introduces a law regarding the use of seat-belts in cars. This law is motivated by statistics on severe injuries and deaths following motor vehicle accidents. Improved curative services are not an option. Knowledge of the determinants for several traffic injuries suggests a protective effect of seat belt use, and other preventative measures, such as the introduction of speed limits, installation of traffic signs and lights, and surveillance, among others. In collaboration with other sectors (such as the ministries of transport, justice and finance), a joint campaign is started for implementation of this law. Car manufacturers must be involved, for example in the development and installation of more comfortable and easier to use devices, and the public must be educated and encouraged (through mass media campaigns) to make use of them. In such a way, all relevant actors are involved in the process leading to and following the decision. In such an approach the probability that people will comply with the preferred decision of protecting car passengers by the correct use of seat-belts is maximised. In a parallel effort, the campaigns may be directed towards improving and promoting the use of public transport, and the use of other transportation means.

This example already suggests some core elements for successful implementation of decisions in environmental health policy, namely:

- The need for information (evidence).
- A focus on determinants rather than on health outcomes.
- The collaboration of different sectors involved in the particular problem.
- The involvement of all relevant stakeholders in society.
- The creation of a supportive environment.

Of these, the need for solid information, relevant and available to all parties, is fundamental. This relates to information on the problem itself as well as information to evaluate the proposed interventions aimed at addressing the problem. In addition, this information must be used in a joint effort of actors in all sectors relevant to the problem concerned. This includes the involvement of those who eventually will receive the health benefits of the decision, namely the community.

In order to elaborate on these core requirements, it is useful to consider a series of real-world examples, from different social and environmental contexts, and reflecting different approaches applied with different degrees of success. These include examples of actions (not necessarily decisions as such) which have improved people's environment, health and their lives as a whole. Of most interest, perhaps, are examples at the local level, because it is at this level that partnerships with, and involvement of, communities can be strongest, and where people can contribute, even on an individual level, to the decisions that affect them.

The first examples show the importance of creating supportive environments by the empowerment of people. Women and children in particular are often relatively disadvantaged in many developing countries, both in rural and urban areas.

- *Example 1. Empowerment of women in a farming area in Zambia.* In a rural area in Zambia the main crop produced was maize, the income from the sale of which was usually kept by men, with little benefit for the women and children, although they contributed considerably to its production. An intersectoral project of several government agencies and womens groups was implemented, which encouraged women to grow vegetables for sale and home consumption. A rural banking service for women was also introduced in the area. The result was the empowerment of women who were enabled to use their own skills for the family's benefit, ensuring a better food supply and thereby improving the nutritional status of their children (Haglund *et al.*, 1996).

- *Example 2. The rights of women in Belo-Horizonte, Brazil*. Profavela is the common term used for a law that recognises the rights of squatter settlements. Its enactment was due in great part to the strength of local community organisations. New legislation introduced in Belo-Horizonte paid special attention to the rights and needs of women. Women were recognised as the cohesive force that keep families together in low income settlements, and since few couples are officially married, property title deeds are preferentially given to women (United Nations, 1996a).

In the following example, empowerment of an ethnic/social minority group provided by government actions created a mutually beneficial situation.

- *Example 3. Waste management in Cairo*. In some cities in developing countries, up to 50 per cent of all the rubbish generated is not collected, but is left to accumulate in the streets where it poses a health hazard. In Cairo, the Zabbaleen people have followed a centuries old tradition of collecting and sorting rubbish found in the city streets. Recently, authorities have turned what was previously a tolerated activity into one which is positively encouraged. This decision proved to be mutually beneficial: the city's waste disposal system has improved and the status and living standards of the Zabbaleen was enhanced. Some 50 recycling and manufacturing businesses have been developed, with non-governmental organisations (NGOs) helping through the provision of basic equipment, training and seed funding (Buckley, 1996).

As the next example shows, having a clear vision of the local environment in which people live, work and recreate, is essential in order to mobilise people to take control of their environment and health. This helps promote community involvement and participation of all those concerned, and collaboration between all sectors which have a role to play in the health of people.

- *Example 4. The importance of a "vision"*. Kuching prides itself on being a "Healthy City" and is recognised as the cleanest and most beautiful in Malaysia. This achievement has, to a large extent, come as a result of pursuing a clear and agreed vision. The city's dream is "*a well-planned, vibrant, landscaped garden city, endowed with a rich artistic, scientific and educational culture. A bustling city with a flourishing and resilient industrial economy, yet clean and unpolluted. A safe city, offering a standard of living affordable by all its citizens. A city managed efficiently and enjoying state-of-the-art communication, information and mass transport technology and providing ready access to services, utilities and recreation areas. A city that is dynamic and attentive to its people's needs and constitutional rights"* (Buckley, 1996).

The importance of having this sort of vision as a guide to action and as a goal for efforts was emphasised in the Habitat II workshop on "Best Practices in Improving the Living Environment", organised by the United Nations in Dubai in 1996 (United Nations, 1996a). As this workshop also showed, however, a vision alone is not sufficient; it cannot be a substitute for decisions and action. The next example shows that the political will to implement the community's vision must also exist.

- *Example 5. Community mobilisation.* From being renowned as the worst polluted city in the USA in 1969, Chattanooga came to be recognised as one of the nation's best success stories. What went right? An initial success in improving air pollution helped to mobilise the community behind a vision to become an "Environmental City". Collaboration between the government, industry and the community generated the required political will, funding and participation to develop strategies to solve existing problems, including housing, transport, recycling and neighbourhood revitalisation. The city has since been called a "living laboratory" for sustainable development (United Nations, 1996b).

Partnerships between the government, communities, the business sector and other important stakeholders are also crucial in laying the foundations for collaboration and success.

- *Example 6. Creating partnerships for action.* Rapid urbanisation in Dar es Salam has caused deterioration of environmental conditions. Environmental hazards include, among others, uncollected solid waste, incomplete incineration of refuse, poorly managed dump sites, and an increased number of unplanned settlements. In 1992 a consultation was held in the city with the purpose of establishing procedures and setting priorities in relation to the "Sustainable Cities" programme of the United Nations Centre for Human Settlements (Habitat). The consultation (with participation of persons from the community, private and government organisations) served to clarify priority urban issues, to establish intersectoral working groups and to establish a multidisciplinary technical support unit. As a result of this work the municipal government, in collaboration with the public and private sectors, began to work on the priority issues identified, with an explicit emphasis on sustainable urban development. The approach has succeeded in widening the basis for participation in the decision-making process and in mobilising a wealth of local resources through partnerships (Bartone *et al.,* 1994; United Nations, 1996a).

- *Example 7. Establishing partnerships and working groups.* Leicester was designated as the UK's first "Environment City" in 1990 because of its record for good environmental practice. Part of the success has been due to an approach based on integrated actions rather than looking at single issues. In addition, the need to identify solutions was stressed rather than just identifying the problems. The approach was to look for partnership rather than confrontation. Local promotional campaigns keep community members involved on a continuous basis. Several working groups were formed bringing together representatives of the community, decision-makers, experts and representatives of the business community, to look at specific areas such as transport, energy and the social environment (United Nations, 1996a; Darlow and Newby, 1997).

This process seems simple and direct. However, partnerships are not without problems. Participation of stakeholders is limited and self-selected. Some partners may feel intimidated in the setting of an "expert group" and will not participate (e.g. community members). Experts may also become frustrated and stop attending if the group disregards what these experts perceive as relevant and important (Darlow and Newby, 1997). Often, there are social and institutional barriers which impede the participation of individuals and community groups (Lawrence, 1996). Evidence, based on solid data and demonstrated to decision-makers is vital to the process of policy and decision-making, as illustrated in the following example.

- *Example 8. Demonstrating the evidence to decision-makers.* In 1990, Sweden introduced a law to limit blood alcohol concentration to 0.2 g l^{-1} for driving a motor vehicle. The new limit was introduced after demonstrating to decision-makers that, despite the popular belief that two beers were sufficient to exceed the 0.5 g l^{-1} limit, a person could drink enough alcohol to feel its effect (a drink before dinner, half a bottle of wine with a meal and a brandy afterwards) but still be under the limit and, thus, be legally able to drive under the influence of alcohol (Haglund *et al.,* 1996).

Focusing on the determinants of health requires long term planning and commitment, and needs strong political will. It has been argued that politicians are more concerned about immediate problems with short term goals (Hunt, 1993), but there are many examples of well-planned long-term projects.

- *Example 9. Public transport in a developing country city.* Curitiba is a city in Brazil which is known for its good "city management". One example is an innovative programme for public transport. Curitiba has more cars per capita than any other city, except for Brasilia, yet it has very few traffic jams. The reason is that 75 per cent of commuters use its public transport system. This was achieved by the introduction of special "busways" and specially designed bus terminals to allow for easy transfer to other routes. One single fare is paid for all journeys within the city limits. In summary, the public transport system is fast, efficient and affordable (Buckey, 1996).

Community commitment is an essential ingredient for success. Even torture for polluters seem to have failed Edward II of England in the example presented at the beginning of this chapter (at least, as evidenced by the fact that his successors had to continue dealing with the problem). More recently, lack of community commitment was one of the reasons for a failed air pollution control mechanism set up in Mexico City.

- *Example 10. Regulation without community commitment.* Mexico City is one of the largest and most (air) polluted cities in the world. Critical air pollutants are ozone, lead, carbon monoxide and fine particulate matter. By 1991, studies had indicated that fine particulates could be causing 12,500 extra deaths and 11.2 million lost days of work per year due to respiratory illness. Ozone was estimated to be responsible for 9.6 million lost work days per year, also due to respiratory illness. Excessive lead exposure was estimated to affect the development of about 140,000 children and cause hypertension in 46,000 adults. Total economic damages were conservatively estimated at US$ 1,500 million per year. An emergency air pollution control programme launched earlier, in 1989, had adopted tight motor vehicle emission standards, vehicle inspections and a rotating one day per week driving ban. However, this regulative approach lacked community commitment and failed. Many drivers bought a second car which, in many cases, was older. The regulation therefore increased the cost of its administration and air pollution (Bartone *et al.*, 1994).

Political boundaries often make control difficult. Transboundary air pollution (between countries) is a well known and documented problem (WHO, 1992). Boundary conflicts may also be a problem in pollution control at the local level.

- *Example 11. Overcoming government boundary conflicts.* Air pollution control in Mexico City is made more difficult because the problem is regional in scope. Air pollution originates in, and affects, the entire Valley of Mexico. Many federal, state and municipal agencies have a say in policy-making, and common actions by different jurisdictional areas (e.g. the implementation of preventative measures) are not easily achieved. To help solve these conflicts, the government created a commission for the prevention and control of environmental pollution in the metropolitan area of the Valley of Mexico, with the role of setting up prevention and control strategies for all aspects of environmental pollution, including air. This committee is now able to define and co-ordinate policies at all levels of government (Bartone *et al.*, 1994).

These case studies illustrate, with differing degrees of difficulty and success, some of the new ways of acting together at the local level. This approach includes efforts to enable and empower local authorities, to improve and use the local "knowledge base", and to build on and encourage the commitment of local people (United Nations, 1996a). The knowledge base is a crucial element — without it, local actions are likely to be poorly informed and inappropriate and, in many cases, will lack the commitment and conviction of the people they are meant to serve. The work described in the following sections and chapters thus concentrates on the question of how to develop and use this local information for decision-making in environmental health.

2.4   Difficulties and uncertainties in the decision-making process

The decision-making process is far from simple, and one in which numerous conflicts and uncertainties arise. One of the basic conflicts derives from the inexact nature of the process: while the public and politicians tend to expect rapid and clear-cut solutions, many problems are often complex and poorly understood, and the scientific evidence is conflicting (Neutra and Trichopoulos, 1993). As Steensberg (1989) stated, there is no definable boundary between what is safe or hazardous, but rather a zone of uncertainty. In many cases, therefore, it is only possible to talk in terms of the probability of an effect being produced. Given the limited public understanding of statistical probabilities and the concepts of risk, such language is not always appropriate or readily accepted (Jardine and Hrudey, 1998).

Decision-making is also bounded by a number of other constraints. Amongst these are problems of data availability and quality, and problems with the analysis and application of findings aimed at determining potential health impacts. Other constraints include uncertainties due to gaps, inconsistencies and errors in many of the data used; inadequate control for all possible confounders; poor quantification of the extent to which prevention can be achieved; extrapolating from evidence derived at high doses to determine risk at lower doses; extrapolating from data derived from animal evidence to determine human risk; extrapolating from past or current data to future health impacts; the need to allow for variations in individual susceptibility; the effects of combinations of exposures and multiple routes of exposure; the unreliability of many of the models used, and the difficulties of model verification; difficulties in defining and valuing intangibles such as quality of life.

Setting clear guidelines to facilitate the decision-making process is therefore not a simple endeavour. All these issues are subject to interpretation, and even experts are likely to disagree regarding both the weight to allocate to each and the conclusions to which they point.

Most decisions involve, and impinge on, a wide range of stakeholders and actors (Whitehead, 1993; Briggs *et al.*, 1998). These typically include scientists, who may be involved in the initial research which identified the problem, and in helping to devise solutions; business and industry, which may be implicated in the cause of the problem and may be partly responsible for implementing and financing solutions; planners, who may be involved in translating general policies into local action, and in monitoring implementation; the media, which may be involved in raising awareness about the problem and act as an unofficial watchdog on the actions taken; politicians, who are charged with making the decisions; and the public, who in the end must accept, pay for and live with the results of the decisions made. Each of these groups is likely to have different agendas. Each will also be moulded by a wide range of economic, professional, political and bureaucratic pressures. Consensus about the levels of risk involved, or about the relative merits of different policy actions, is therefore difficult to achieve (McMichael, 1991).

The need to involve the various actors and stakeholders at all stages in the decision process should not be treated lightly. Some questions, for example, are unanswerable in strictly scientific terms because of gaps in our knowledge. In these cases, a dialogue with the community is essential in order to reach a mutually agreeable solution. Science can provide guidance but not provide all the answers. An open and participatory approach is more likely to make the results more credible and acceptable, to provide time for the community to consider in advance the technical concepts and limitations and range of outcomes, and thus to allow decisions to be taken and implemented more effectively and speedily (Ozinoff and Boden, 1987). It is recognised, however, that the political process must support a participatory approach. In certain societies, civic organisations have remained weak, not formally recognised, repressive or non-existent. In such cases, an open participatory process is unlikely to be undertaken satisfactorily.

In this context, de Koning (1987) noted five characteristics of an effective standard-setting process which can be applied generally to decision-making in the area of environmental health:

- Involve the major parties in the community, including politicians, citizen groups, industrial leaders and health officials. This should stimulate debate encompassing differing perspectives and values, leading to some compromises being made in both goals and methods, thus ensuring broad support in the society at large.

- Provide a mechanism through which technical and policy analysis can be generated, distributed and criticised.
- Provide a mechanism whereby the results of analyses can be presented to policy-makers and the other centres of interest in the society, to inform these groups of the costs, benefits, and impact of the proposals under consideration.
- Provide a mechanism for conflicting interests to be heard and discussed in a controlled manner, so that divergent opinions in the society can be aired and, as far as possible, accommodated in the implementation of the proposal.
- Provide a mechanism whereby the society can reach a decision and take useful action, even though such action may be less than what is objectively ideal.

2.5   Conclusions

Decision-making requires the availability of better information and knowledge on the links between environment and health, but epidemiological research results are seldom definitive or conclusive (Omenn, 1993). However, it is inadviseable to delay while this information and knowledge is gathered, because while waiting for the information the problem continues and those affected have a right to know and to be protected (Sandman, 1991). It is necessary to be prepared, therefore, to act with the data and methods available.

Other chapters describe methods and tools to aid the decision-making process. The purpose of these is to help extract more information, more quickly, out of the data that already exist — and where adequate data are not available, to collect them speedily. The aim is to improve the utility of the information gained by providing results in a form directly usable by the decision-maker. As part of this purpose, the clear need is to encourage epidemiologists and other scientists to work more closely with decision-makers and each other, and for all three groups to interact more openly with the public and other stakeholders concerned.

2.6   References

Ashton, J. 1992 The origins of healthy cities. In: J. Ashton [Ed.] *Healthy Cities*. Open University Press, London.

Bartone, C., Bernstein, J., Leitman, J. and Eigen, J. 1994 *Toward Environmental Strategies for Cities — Policy Considerations for Urban Environmental Management in Developing Countries*. The World Bank, Washington.

Bradley, D. 1994 Health, environment, and tropical development. In: B. Cartledge [Ed.] *Health and the Environment*. Oxford University Press, Oxford.

Briggs, D.J., Stern, R. and Tinker, T.L. [Eds] 1998 *Environmental Health for All. Risk Assessment and Risk Communication for National Environmental Health Action Plans*. NATO Science Series 2, Environmental Security - Vol. 49, Kluwer, Dordrecht.

Brown, V.A., Ritchie, J.A. and Rotem, A. 1992 Health promotion and environmental management: a partnership for the future. *Health Promotion International*, **7**, 219–30.

Buckley, R. [Ed.] 1996 *Understanding Global Issues*. Understanding Global Issues Limited, Cheltenham.

Darlow, A. and Newby, L. 1997 Partnerships: panacea or pitfall? Experience in Leicester environment city. *Local Environment*, **2**, 73–81.

de Koning, H.W. 1987 *Setting Environmental Standards. Guidelines for Decision-making*. World Health Organization, Geneva.

Graham, D.J. and Wiener, J.B. 1995 Confronting risk trade-offs. In: D.J. Graham and J.B. Wiener [Eds] *Risk Versus Risk — Trade-offs in Protecting Health and the Environment*. Harvard University Press, Cambridge Massachusetts.

Haglund, J.A., Finer, D., Tillgren, P. and Pettersson, B. [Eds] 1996 *Creating Supportive Environments for Health*. World Health Organization, Geneva.

Hunt, S.M. 1993 The relationship between research and policy. In: J.K. Davies and M.P. Kely [Eds] *Healthy Cities*. Routledge, London.

Jardine, C.G. and Hrudey, S.E. 1998 What is risk? In: D.J. Briggs, R. Stern and T.L. Tinker [Eds] *Environmental Health for All. Risk Assessment and Risk Communication for National Environmental Health Action Plans*. NATO Science Series. 2. Environmental Security Vol. 49, Kluwer Academic Publishers, Dordrecht, 205–11.

Lalonde, M. [Ed.] 1974 *A New Perspective on the Health of Canadians — a Working Document*. Health and Welfare, Canada.

Lawrence, R.J. 1996 Wanted: designs for health in the urban environment. *World Health Forum*, **47**, 363–6.

McKeown, T. 1976 *The Role of Medicine — Dream, Mirage or Nemesis?* The Nuffield Provincial Hospitals Trust, UK.

McMichael, A J. 1993 *Planetary Overload — Global environmental change and the Health of the Human Species*. Cambridge University Press, Cambridge.

McMichael, A.J. 1991 Setting environmental exposure standards: current concepts and controversies. *International Journal of Environmental Health Research*, **1**, 2–13.

Neutra, R.R. and Trichopoulos, D. 1993 The place of epidemiology in environmental decisions: Needed support for the development of risk assessment policy. *Environmental Health Perspectives*, **101**(S4), 67–9.

O'Neill, M. 1993 Building bridges between knowledge and action. In: J.K. Davies and M.P. Kely [Eds] *Healthy Cities*. Routledge, London.

Omenn, G. 1993 The role of environmental epidemiology in public policy. *Annals of Epidemiology* **3**, 319–22.

Ozinoff, D. and Boden, L.I. 1987 Truth and consequences: health agency responses to environmental health problems. *Science, Technology and Human Values*, **12**, 70–7.

Sandman, P. 1991 Emerging communication responsibilities of epidemiologists. *Journal of Clinical Epidemiology,* **44**, 41S–50S.

Steensberg, J. 1989 *Environmental Decision Making: The Politics of Disease Prevention.* Almqvist and Wiksell International, Copenhagen.

Sundin, J. 1990 Environmental and other factors in health improvement explaining increased survival rates in 19th century Sweden. In: E. Nordberg and D. Finer [Eds] *Society, Environment and Health in Low-Income Countries.* Karolinska Institute, Sweden.

United Nations 1993 *Agenda 21: Programme of Action for Sustainable Development.* United Nations, New York.

United Nations 1996a *Dubai International Conference for Habitat II on Best Practices in Improving the Living Environment.* United Nations, Nairobi.

United Nations 1996b *Habitat Debate.* United Nations Centre for Human Settlements, Nairobi.

Warford, J.J. 1995 Environment, health, and sustainable development: the role of economic instruments and policies. *Bulletin of the World Health Organization,* **73**, 387–95.

Warner, D.M., Holloway, D.C. and Grazier, K.L. 1984 *Decision Making and Control for Health Administration.* Health Administration Press, Ann Arbor.

Whitehead, M. 1993 The ownership of research. In: J.K. Davies and M.P. Kelly [Eds] *Healthy Cities.* Routledge, London.

WHO 1978 *Report of the International Conference on Primary Health Care.* WHO Health for All Series No. 1, World Health Organization, Geneva.

WHO 1992 *Our Planet, Our Health — Report of the WHO Commission on Health and Environment.* World Health Organization, Geneva.

WHO 1996 *The World Health Report 1996: Fighting Disease, Fostering Development.* World Health Organization, Geneva.

Wilson, R. and Spengler, J. 1996 *Particles in Our Air — Concentrations and Health Effects.* Harvard University Press, Harvard, Massachusetts.

World Commission on Environment and Development 1987 *Our Common Future.* Oxford University Press, Oxford.

Chapter 3[*]

THE NEED FOR INFORMATION: ENVIRONMENTAL HEALTH INDICATORS

3.1 Introduction

It has been recognised that in many parts of the developing world the burden of disease attributed to environmental factors is large (WHO, 1997). Even in the developed world (and focusing for simplicity on the physical environment) new pollutants are emerging which pose threats to human health, and for which the health burden estimates are unknown or hard to measure. Against this background, there is clearly an urgent need for action to reduce the environmental health burden. This can be achieved, for example, through:

- Technological innovation to develop new, cleaner and more sustainable methods of production.
- Demand control to reduce the pressures from consumption and resource use.
- Environmental improvement to reduce the hazards involved, especially in those areas where human exposure may occur.
- Education and awareness raising to help individuals better appreciate the environmental risks to which they are exposed, and the personal opportunities which exist for risk avoidance and reduction.
- Therapeutic interventions to minimise the health impact on those already affected.

For any given environmental health problem, actions need to be taken through all the measures specified above. Certainly, technological innovations are likely to have a sustained, longer-term impact, but in the short-term public education and even therapeutic actions are also needed. All of these actions are potentially costly and therefore they all depend on the availability of reliable information. Information may thus be needed for the following (Briggs, 1995):

- To help identify and prioritise the problems which exist.
- To inform the numerous groups of stakeholders involved.
- To provide a rational framework for discussion and debate.
- To define, evaluate and compare the actions which might be taken.

[*]   *This chapter was prepared by C. Corvalán, D. Briggs and T. Kjellström*

- To monitor the effects of these actions.
- To help specify safe limits and environmental guidelines and standards.
- To guide the research and development needed for the future.

The need for information to support policy and action in environmental health has been introduced in Chapter 2. This chapter focuses in more detail on the development of indicators suitable for decision-making. It takes on an epidemiological approach to understand the development–environment–health linkages and concentrates primarily on the technical aspects of obtaining usable and relevant environmental health information.

3.2 Indicators of development, environment and health

The term "indicator" is derived from the Latin *indicare*, meaning to announce or point out. Indicators represent more than the raw data on which they are based. They provide a means of giving the data added value by converting them into information of direct use to the decision-maker. Indicators are thus a crucial link in the data–decision-making chain: measurements produce raw data; data are aggregated and summarised to provide statistics; statistics are analysed and re-expressed in the form of indicators; and indicators are then fed into the decision-making process (Wills and Briggs, 1995). As such, an environmental health indicator can be seen as a measure which summarises, in easily understandable and relevant terms, some aspect of the relationship between the environment and health which is amenable to action. It is a way, in other words, of expressing scientific knowledge about the linkage between environment and health in a form which can help decision-makers to make better informed and more appropriate choices.

Environmental health indicators have the potential to contribute to improved environmental health management and policy. They are, however, of particular value in countries in which problems of access to natural resources remain, and in which issues of environmental pollution have traditionally taken second place to demands for economic development. Many of these countries are also confronted with hazards and diseases associated with poverty and lack of development (Environmental Research, 1993). In many countries problems of resource depletion, desertification and environmental pollution are rising. At the same time, populations are undergoing rapid expansion, particularly in urban centres, and these changes are in turn becoming an important driving force behind health and environment problems (Stephens, 1995; Harpham and Blue, 1997). In recent years, awareness has been growing of the association between economic growth and environmental protection (World Bank, 1992; United Nations, 1993) and, in many countries, strategies for sustainable development aimed at both preserving the environment and enhancing quality of life are being implemented (e.g.

Projecto Estado de la Nacion, 1995; Environmental Health Commission, 1997). If decision-makers are to take the actions needed to prevent irreversible and costly health and environmental damage, they urgently need reliable and relevant information on levels of environmental pollution and their links with human health.

The concept of indicators is far from new. The use of indicators has a long history, for example in economics (e.g. indicators such as Gross National Product (GNP) and the unemployment rate), resource management (e.g. indicators of land suitability) and ecology (e.g. the use of indicator species and of ecosystem health) (Rapport, 1992). In recent years, however, there has been a marked growth in interest in the use of indicators in many other fields. The use of social indicators (e.g. of deprivation, poverty) is now widely accepted (e.g. Jarman, 1983; Carstairs and Morris, 1989; UNDP, 1997), while performance indicators are being used increasingly to monitor the activities of industry and the public services. Indicators have also become well-established in the fields of both environment and health (e.g. UNEP, 1993; WHO, 1993a).

There are four main categories of indicators in use that are considered relevant in the context of development, environment and health. These are sustainable development indicators, environmental indicators, health indicators and environmental health indicators. While there are important overlaps among these, the focus of this chapter is on the indicators which can contribute usefully to environmental health policies.

3.2.1   Sustainable development indicators
One of the most important stimuli for indicator development in the areas of environment and health has been the emergence of sustainable development as a guiding principle for policy, and the adoption in 1992 of Agenda 21 at the United Nations Conference on Environment and Development (UNCED) (see Chapter 2). Countries and international governmental and non-governmental organisations (NGOs) were called upon to develop the concept of indicators of sustainable development. The Statistical Office of the United Nations was given a special role to support this work and to promote the increasing use of such indicators. National programmes for indicator development have thus been set up in many countries to support environmental policy and State of the Environment reporting (e.g. Environment Canada, 1991; Adriaanse, 1993). The adoption of Local Agenda 21 has similarly encouraged the establishment of sustainability indicators by local governments and city authorities (e.g. Gosselin et al., 1993; Sustainable Seattle, 1993; Local Government Management Board, 1994). Internationally, several organisations have attempted to construct core sets of indicators to monitor

global environmental trends (e.g. OECD, 1993, 1997; UNEP/RIVM, 1994; World Bank, 1994; World Resources Institute, 1995; Worldwide Fund for Nature and New Economics Foundation, 1994).

The United Nations has recently listed 130 sustainable development indicators to be tested in countries (United Nations, 1996). Many of these indicators, however, do not reflect the sustainability aspect they wish to measure. Economic performance indicators, such as GNP or the annual GNP increase, tell us nothing about the ability of future generations to sustain development. In fact, it could be speculated that a high GNP today may be the direct cause of a lowered GNP tomorrow, if natural resources are depleted and the high current GNP has been created at the expense of the community's future productivity. Although the concept of sustainable development has, to some extent, been adopted by politicians to refer to short-term economic goals, economic performance in itself is not the ultimate aim of sustainable development. Instead, long-term human health and welfare, biodiversity protection and global ecosystem health are the key objectives of sustainable development (Gouzee *et al.,* 1995). Most environmental indicators (e.g. air quality) or health indicators (e.g. life expectancy) provide no information about sustainability as such, but they are at least essential elements of community well-being. Some environment and health indicators can also be interpreted more directly in relation to sustainability. For example, an indicator of soil quality or soil stability could be interpreted as directly linked to future agricultural productivity and the ability of future generations to meet their needs. Similarly, a health indicator, such as the occurrence of infectious disease in a community, could be interpreted in relation to likely health problems in the future.

Attempts have also been made to assess other aspects of development, for example human development. An example of this is the human development index (UNDP, 1990). More recently, other measures of human development have been introduced, such as the human poverty index (UNDP, 1997).

3.2.2   Environmental indicators

Environmental indicators have been described as "*a measurement, statistic or value that provides a proximate gauge or evidence of the effects of environmental management programs or the state or condition of the environment*" (US EPA, 1994). In recent years, several programmes have been established to monitor the environment for health-related purposes, for example the Global Environment Monitoring System (GEMS) for air (UNEP/WHO, 1993; see also WHO, 1987), water (WHO, 1991) and food (WHO, 1990). Nevertheless, issues relating to health are just a few of the many reasons for collecting environmental indicators. Other reasons include the impact of environmental pollution on agriculture, forests, rivers and lakes. Thus, the collection of data on air pollution emissions and concentrations, organic and inorganic water pollution, stratospheric ozone, natural resources, waste production, climate change, etc., is not performed specifically for health related purposes. In the context of human health it is mostly the degree of exposure of humans to potential health risks that is of concern, and consequently the human health impact of contaminants (and other risk factors) in the environment.

The difficulty with environmental indicators is that the presence of pollutants in the environment does not translate automatically into health outcomes. Similarly, the incidence of many environmentally-related diseases cannot be easily traced back to specific environmental exposures. Only individual-level epidemiological studies are able to establish reliable links between exposures and health outcomes. Such studies, however, cannot on their own provide the information needed to support action and policy, and defeat the purpose of using easily collected or available statistics to derive, quickly and cost-effectively, environmental health indicators.

3.2.3   Health indicators

Health indicators have been used extensively to monitor the health of populations. The "Health for All" policy, for example, involves monitoring progress towards a minimum health level for all persons by the year 2000 and provides numerous examples of health indicators on a global scale. The information gained from monitoring is used for evaluation, i.e. the continuous follow-up of activities to ensure that they are proceeding according to plan, so that if anything goes wrong, immediate corrective measures can be taken (WHO, 1993a). The health–environment link is also a prominent part of the "Health for All" process. Important environmental health issues, such as access to water and sanitation, acute and chronic exposures to chemicals, population exposures to unacceptable levels of contaminated air, housing issues (as well as broader environmental issues with a less direct link to health, such as loss of biodiversity, deforestation, soil degradation and global warming) are all addressed in the publication *Implementation of the Global Strategy for Health for All by the Year 2000* (WHO, 1993b).

Health indicators are usually defined in terms of health outcomes of interest. The Swedish Environmental Protection Agency has compiled a tentative list of environment-related diseases (SEPA, 1993) which can be used for this purpose. This list includes certain cancers (especially lung and skin, particularly in children); respiratory disease (chronic bronchitis, pulmonary emphysema, bronchial asthma, hyper-reactivity); allergic diseases (atopic allergies and symptoms occurring in connection with atopic diseases, namely asthma, hay fever, conjuctival catarrh and eczema); cardiovascular diseases; effects on reproduction (miscarriage, late intrauterine death, neonatal and perinatal death, low birth weight, various malformations and chromosome abnormalities); and diseases of the nervous system (organic psychosyndromes and dementia (Alzheimer's disease), Parkinson's disease, amyotrophic lateral sclerosis, peripheral nervous disease in combination with polyneuropathy). Not all cases of these diseases are due to environmental exposures and not all environment-related diseases are included in this list. For example, certain infectious diseases would be prominent environment-related diseases in less developed countries. Nevertheless, these diseases do provide a means of monitoring and assessing the health outcome of a wide range of environmental exposures.

The term "public health surveillance" is used to describe the collection, analysis and interpretation of data on specific health events, for the purpose of prevention and control (Thacker *et al.*, 1996). Surveillance in environmental health extends this concept by including surveillance of hazards and exposures (Hertz-Picciotto, 1996; Thacker *et al.*, 1996). The term "sentinel health event" has been applied to cases of disease that, in a particular situation, appear out of the ordinary, and which can be potentially linked to an external factor, for example infant or maternal deaths as indicators of the adequacy or quality of prenatal or maternal health care. The concept of sentinel health events is especially appropriate in relation to occupational health, and currently more than 50 conditions are considered as occupational sentinel health events (e.g. asbestosis and mesotheliomas as indicators of asbestos exposure) (Mullan and Murthy, 1991). A preliminary list of environmentally-related sentinel health events has also been devised (Rothwell *et al.*, 1991). In practice, however, there are few diseases which can be used as sentinels of environmental exposures.

3.2.4   Environmental health indicators

Environmental health is concerned not with the health of the environment *per se*, but with the ways in which certain environmental factors can influence or directly affect human health (in either a positive or negative way). An environmental health indicator can thus be defined as:

> "*an expression of the link between environment and health, targeted at an issue of specific policy or management concern and presented in a form which facilitates interpretation for effective decision-making*".

Several aspects of this definition are worthy of emphasis. The first is that an environmental health indicator embodies a linkage between the environment and health. As such it is more than either an environmental indicator or a health indicator. Environmental indicators represent indicators which describe the environment without any explicit or direct implications for health. The vast majority of environmental indicators so far developed are of this type, for example indicators of atmospheric emissions, surface water quality, designated areas or threatened wildlife species. Health indicators are indicators which describe the status of, or trends in, health without any direct reference to the environment. Again, the majority of health indicators so far developed are of this type; examples include simple measures of life expectancy, or cause-specific mortality rates where no attempt has been made to estimate those health outcomes attributable to the environment.

Given knowledge of the relationship between specific environmental exposures and health effects, however, both environmental indicators and health indicators can be converted into environmental health indicators. An environmental health indicator is thus a measure which indicates the health outcome due to exposure to an environmental hazard. As such, it is based upon the application of a known or postulated environmental-exposure health–effect relationship. In this respect, two general types of environmental health indicators can be distinguished:

- An *exposure-based indicator* projects forward from some knowledge about an environmental hazard to give an estimated measure of risk. Such indicators can be conceived as the combination of an environmental indicator with a known environment–health relationship (e.g. the estimated health impact, such as respiratory disease, from known levels of air pollution).
- An *effect-based indicator* projects backwards from the health outcome to give an indication of the environmental cause (i.e. the environmentally attributable health outcome, such as the proportion of current diarrhoea death rates which can be attributed to poor water quality).

Within the context of environmental health, the word "environment" is understood to comprise all that which is external to the human host, including physical, biological and social aspects, any or all of which can influence the health status of populations (Last, 1995). The environment, therefore, encompasses not only the general environment to which everyone is exposed, but also specific environments, such as the workplace and the domestic environment, where people spend a significant proportion of their time. Further, one must also include among environmental health hazards not only the immediate biological, chemical or physical factors that affect health, but also the underlying social, economic and technical conditions that give rise to (and modify) environmental health problems. An indicator which purely describes the state of the environment with no obvious link to health impacts of the environment could not be considered an environmental health indicator. In the same vein, a pure health status indicator with no obvious linkage to environmental causation of health deterioration (or health improvement), could not be considered an environmental health indicator.

Figure 3.1 provides a graphic description of the relationship between the three related arenas of environment, health and environmental health. The

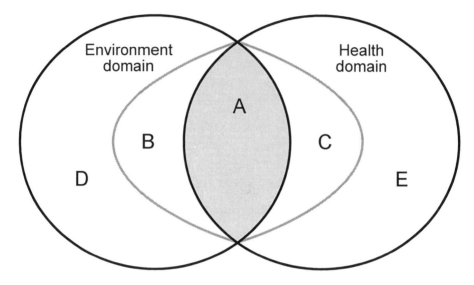

Figure 3.1 The scope of environmental health indicators. **A.** Environmental health indicators; **B.** Environmental indicators indicating potential human health impacts; **C.** Health indicators with unknown but possible environmental cause; **D.** Well defined environmental indicators; **E.** Well defined health indicators

area relating to environmental health indicators (area A) is formed by the area of intersection (or linkage) between the environment and health. This is the area in which known (or suspected) environmental factors are associated with known (or suspected) health outcomes; for example the effects of severe air pollution on respiratory disease in children, or of poor sanitation on gastro-intestinal disease. Area B represents the area in which the environment, while not directly influencing human health, may nevertheless contribute more tenuously to health outcome. Examples of this include deforestation and desertification, which do not have a well-defined, direct or immediate linkage with health; or environmental exposures which we have not yet identified as hazardous to health. Area C represents health outcomes (e.g. diseases such as certain cancers) with unknown but possible environmental causes. Areas D and E represent those areas of environment and health, respectively, wholly outside the realm of environmental health, i.e. where there is no apparent link between environment and health. For the sake of simplicity, the schema presented in Figure 3.1 deliberately excludes factors, such as economic and social conditions, which may affect the environment and health (House *et al.,* 1988) but which may act as modifiers of the health effects resulting from the environment. Poverty, for example, may exacerbate the health effects of exposures to environmental pollution both by increasing susceptibility of the population (as a result, for example, of inadequate nutrition) and by reducing access to early health treatment (e.g. Ostro, 1994).

Indicators can be devised and constructed for each of the areas shown in Figure 3.1. Because reliable environmental health indicators can only be developed where the association between environment and health is clear and strong, the most useful indicators occur in area A. In areas B and C, the link between environment and health is either weaker or less certain. In these areas, therefore, reliable indicators are more difficult to define, and any environmental health indicator will need to be interpreted with particular care; it will rarely be possible to assume that changes in the indicator necessarily reflect the effect of environment on health. Areas D and E are the terrain of explicit and independent environmental indicators and health indicators. Indicators in this area cannot be considered legitimately to be measures of environmental health.

The areas shown in Figure 3.1, however, are not fixed. The boundaries between the various areas may change as our knowledge of the links between environment and health develop. As knowledge improves (e.g. as a result of advances in epidemiological research), therefore, so area A may expand to encompass progressively more of areas B and C. As new theories emerge about potential environment–health effects, so areas B and C may expand into D and E. Equally, new research may disprove assumed relationships, causing a contraction of the area occupied by environmental health. In the process, the meaning and utility of existing indicators may change, and opportunities may develop for the construction of new indicators, aimed at new concerns.

Another important characteristic of an environmental health indicator is its relationship with policy or management. Any environmental health indicator must be useful. To be useful, it must relate to aspects of environmental health which are both of relevance to the decision-maker and, directly or indirectly, amenable to control. Given that the collection of information invokes costs, and that these costs will need to be justified, it will rarely make sense to collect information or try to construct indicators which will not be used in support of policy. This means that most indicators are built around areas of existing policy; the policy imperative creates both the need for indicators and justifies the costs of constructing them. Some of the most valuable uses of indicators, however, are to help identify and assess new policy questions. This means that some indicators need to be developed in advance of a clear and definite policy need. A spectrum of environmental health indicators can thus be identified, reflecting the strengths of their links with policy. At one end are those indicators which have a clear and known use in relation to existing policy or recognised concerns. In the middle are those indicators which are based on less clear policy needs, but which over time may help to guide and direct new policy developments. Beyond, lie those indicators with no apparent policy relevance. Because of the economic considerations involved in indicator development, most attention tends to be devoted to the first of these three categories. However, the uncertainty of present knowledge about environmental health, the length of time it often takes to investigate new problems, and the long latency times which often exist between exposure and health effect, mean that risks are being taken if attention is not focused in this area. Thus, new problems are likely to occur unexpectedly. Consequently, the "precautionary principle" needs to prevail and indicators are needed that give an early warning of new environmental health effects in time to address them before they become severe. In the long term, it is therefore the more prospective indicators (i.e. those at the margins of existing policy) that are often the most important. Unfortunately, these are often the most difficult to justify.

A third aspect of this definition is that environmental health indicators must be expressed in a way which is pertinent to, and understandable by, the decision-makers concerned (Gosselin *et al.*, 1993). In many circumstances, this requires that the indicator be expressed in terms of the health risk associated with a specific environmental hazard, because this provides a universally recognisable "currency" by which to assess and compare different problems. Possibly the most meaningful measure is thus one that provides estimates of the severity and magnitude of the health outcome (e.g. the number of additional deaths, the number of additional hospital admissions, or the number of additional cases of morbidity). In practice, it is often difficult to calculate with any certainty the actual health effect in these terms, because these estimates rely upon having a quantitative understanding of the dose–response relationship. An alternative may be to express the indicator in terms of the number of people "at risk". This can often be estimated from knowledge of the levels of exposure across the population. Often, however, even this may be difficult (e.g. where pollution levels are measured at too few sites to allow estimates of the population exposure). In these situations, the indicator may be expressed simply in terms of environmental concentrations, or some measure of source activity. The further the indicator is removed from the health outcome, the less clearly it expresses the health risk involved, and the more uncertain any interpretations of these risks will be. On the other hand, because policy action, especially preventative action, is often targeted at the source of the pollution, these more remote, source-based indicators may still be very valuable in terms of guiding policy.

All these considerations have important implications for the way in which data is collected and the indicators concerned are constructed and presented. Some of the criteria which help to make good environmental indicators are therefore considered in the next section.

3.2.5   Criteria for indicator development

While indicators are intended to provide a simplification of reality, they are themselves far from simple. Unless this underlying complexity is understood, indicators may end up being developed in relatively fuzzy and ill-defined terms. Gosselin *et al.* (1993), for example, derived from an extended list, a set of 20 indicators for measuring and reporting progress of sustainability. Among these are indicators which are relatively self-explanatory both in terms of what they are meant to indicate and how they should be constructed and measured; for example energy consumption per capita or employment to population ratio. Others, however, are less clearly defined (e.g. public transport use compared with car or major water pollutant emissions); these would need to be further clarified before they were developed. As a result, it is not clear how the indicator should be measured, what data are needed, or how it can be interpreted. As this implies, clear definitions and explanations about every aspect of the indicator to be used are crucial. Poorly conceived or inadequate indicators are likely to be a waste of time and effort, and they are likely to misinform, rather than inform, the users.

It is all too easy, therefore, to propose indicators which do not, in reality, indicate anything — or at least not what the user assumes. Good indicators require careful planning and design. They depend upon an understanding of the questions being addressed, of the way in which they will be used, and of the way in which the systems involved operate. In addition, they need to be formulated and defined very precisely, and they often need to be tested before they can be used.

Fortunately, in recent years much has been learned about the development and use of indicators in a wide range of decision-making areas. On the basis of this experience, a number of criteria have now been established for general indicator selection and construction (e.g. Kreisel, 1984; UNEP/RIVM, 1994; OECD, 1997). These can be further adapted in relation to environmental health. Pastides (1995) takes an epidemiological approach to arrive at two fundamental criteria:

- That the indicator should reflect an underlying causal mechanism.
- That the indicator should be a valid estimate of the causal relationship.

If indicators are to be used to assist decision-making, however, they cannot be judged solely in terms of their scientific validity. Factors such as utility, acceptability and cost of construction also become important. For most purposes it is thus more useful to recognise two fundamental sets of criteria: those relating to their scientific validity and those relating to their relevance and utility. Box 3.1 lists some of the main criteria that can be identified under each of these headings. It is important to recognise that not all these criteria can necessarily be achieved in all circumstances. Problems of data availability, resources and the need for compatibility with previous indicator

---

Box 3.1   Criteria for environmental health indicators

Environmental health indicators should be:

A. Scientifically valid

- Based on a known linkage between environment and health.
- Sensitive to changes in the conditions of interest.
- Consistent and comparable over time and space.
- Robust and unaffected by minor changes in methodology/scale used for their construction.
- Unbiased and representative of the conditions of concern.
- Scientifically credible, so that they cannot be easily challenged in terms of their reliability or validity.
- Based on data of a known and acceptable quality.

B. Politically relevant

- Directly related to a specific question of environmental health concern.
- Related to environmental and/or health conditions which are amenable to action.
- Easily understood and applicable by potential users.
- Available soon after the event or period to which it relates (so that policy decisions are not delayed).
- Based on data that are available at an acceptable cost–benefit ratio.
- Selective, so that they help to prioritise key issues in need of action.
- Acceptable to the stakeholders.

---

series may mean that some have to be sacrificed. Nevertheless, the criteria listed in Box 3.1 provide a useful checklist against which to judge environmental health indicators, and in general good indicators are likely to satisfy most, if not all, of these criteria.

3.3   A conceptual framework for environmental health indicators

3.3.1   The environment–health chain

The link between environment and health operates through the exposure of humans to environmental hazards. These hazards may take many forms — some are wholly natural in origin whereas the majority derive from human activities and interventions. In all cases, however, health effects only arise if humans are exposed, often at a specific place and time, to the hazards which exist.

The environment–health pathway is most clearly seen in the case of exposure to pollution (Lioy, 1990; Sexton *et al.*, 1992, 1994; Pirkle *et al.*, 1995) (Figure 3.2). Most environmental pollutants are the product of human activities. These may be released into the environment in a variety of ways, and

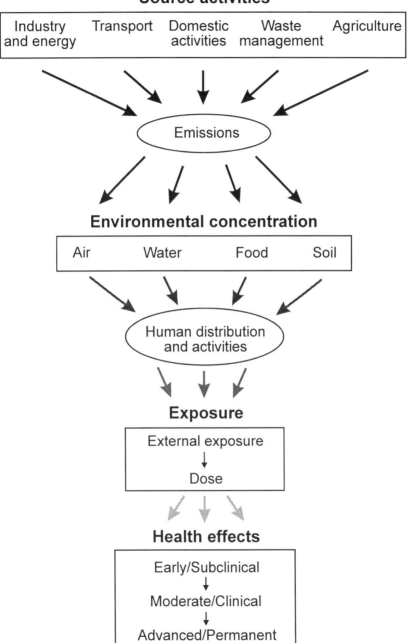

Figure 3.2 The environmental health hazard pathway. Arrows indicate the flow from source activities to health effects (e.g. pollutants). Arrow shading indicates the likely weakening of the impact from source activity to health effects.

may then be dispersed and accumulate in different environmental media (e.g. the air, water, soil, food). Exposure occurs when humans encounter the contaminants within any one of these media. A range of health effects may then occur, from minor sub-clinical effects through illness to death, depending upon the intrinsic harmfulness of the pollutant, the severity of exposure and the susceptibility of the individuals concerned. The whole process is often driven by persistent forces which motivate the creation of the hazard and increase the likelihood of exposure. Thus, population growth, economic development, technological change and (behind these) social organisation and policies may all lie at the root of the problem. Because of its potentially wider impact, it is often here where action needs to be addressed.

3.3.2   The DPSEEA framework

The environment–health chain illustrated by the example of pollution provides a useful organising framework for the development and use of environmental health indicators. However, to make it more generally applicable (e.g. to other forms of environmental hazards), and to set it more firmly within a decision-making context, it needs to be further conceptualised.

Over recent years, a number of attempts have been made to devise conceptual frameworks for indicator development. Of these, the one which has been most widely adopted has been the simple pressure–state–response (PSR) sequence, initially applied by the Organisation for Economic Co-operation and Development (OECD) as a framework for State of the Environment reporting (OECD, 1993, 1997). A slightly modified version is currently in use by the United Nations to develop sustainable development indicators (United Nations, 1996). In many ways, however, the PSR sequence has proved too limiting, and it has more recently been extended to include recognition of the "driving forces" responsible for pressures on the environment, and of the effects which often precede the policy response (e.g. US EPA, 1994). Figure 3.3 further adapts these concepts to provide a specific framework which addresses the driving forces, pressures, state, exposures, effects and actions (i.e. DPSEEA) for the development of environmental health indicators. This framework acts as a valuable guide to designing indicators in a wide range of situations; for example in developing indicators to address a specific environmental hazard (e.g. air pollution) or a specific health problem (e.g. respiratory illness in children), or to describe the whole web of links between environment and health which may occur in a specific area (e.g. a local community). It has also proved useful in describing and analysing the global situation in relation to health, environment and development in a recent report entitled *Health and Environment in Sustainable Development* (WHO, 1997).

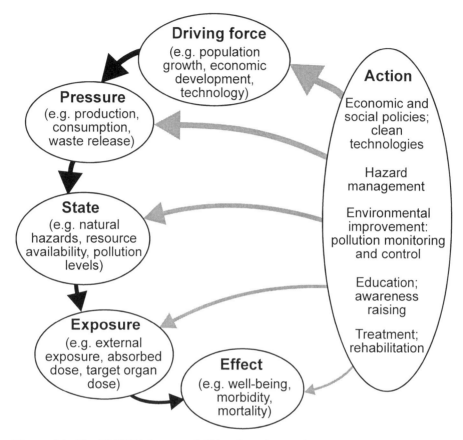

Figure 3.3 The DPSEEA framework (After Corvalán *et al.* 1999)

Driving forces

Within the DPSEEA framework, the driving forces component (D) refers to the factors which motivate and push the environmental processes involved. One of the most important of these is population growth (Canadian Journal of Public Health, 1991; Winkelstein, 1992; McMichael, 1993; Bongaarts, 1994). Almost inevitably, this results in more people being exposed to environmental hazards simply by virtue of the increased number of people living in the areas concerned. More indirectly, it tends to lead to the intensification of human activities within these areas, thereby contributing to environmental damage and resource depletion (Litsios, 1994). In some cases it also results in expansion of human populations into more marginal zones. Here, the inherent instability of the environment may mean that the population is especially vulnerable to environmental hazards, while the environment in turn is especially sensitive to damage. A wide range of other important driving forces also exist, including technological development, economic development and policy intervention (e.g. see Warford, 1995).

Pressures

The driving forces within the DPSEEA model result in the generation of pressures (P) on the environment. These pressures are normally expressed through human occupation or exploitation of the environment. Pressures are thus generated by all sectors of economic activity, including energy production, manufacturing, transport, tourism, mining and agriculture. In each case, pressures arise at all stages in the supply chain — from initial resource extraction, through processing and distribution, to final consumption and waste release.

One of the most important components of these pressures in the context of human health is clearly the release of pollutants into the environment. These releases may occur in a wide variety of ways, and into different environmental media. Energy combustion, for example in vehicles, manufacturing industries, electricity generation and home heating, is one of the most important emission processes, especially to the air. Large quantities of pollutants are also emitted through other processes, such as spillage of chemicals, the deliberate discharge of effluents, dumping of wastes, leaching of agricultural chemicals, etc. Because these activities and processes represent the starting point for environmental emissions they also represent the most effective point of prevention and control. Once in the environment, pollutants may be widely dispersed and may undergo a wide range of secondary transfers. Environmental policy is therefore focused at trying to regulate source activities, or to incorporate in them methods of emission control.

State

In response to these pressures, the state of the environment (S) is often modified. The changes involved may be complex and far-reaching, affecting almost all aspects of the environment and all environmental media. Thus changes occur in the frequency or magnitude of natural hazards (e.g. in flood recurrence intervals or in rates of soil erosion); in the availability and quality of natural resources (e.g. soil fertility, biodiversity); and in levels of environmental pollution (e.g. air quality, water quality). These changes in the state of the environment also operate at markedly different geographic scales. Many changes are intense and localised, and are often concentrated close to the source of pressure (e.g. habitat loss, urban air pollution, contamination of local water supplies). Many others are more widespread, contributing to regional and global environmental change (e.g. desertification, marine pollution, climate change). Because of the complex interactions that characterise the environment, almost all these changes have far-reaching secondary effects.

Exposure

Environmental hazards, however, only pose risks to human well-being when humans are involved. Exposure (E_1) thus refers to the intersection between people and the hazards inherent in the environment. Exposure is rarely an automatic consequence of the existence of a hazard. It requires that people are present both at the place and at the time that the hazard occurs.

The concept of exposure is best developed in relation to pollution. The National Academy of Sciences (1991) defines exposure as: "*an event that occurs when there is contact at a boundary between a human and the environment with a contaminant of a specific concentration for an interval of time*". In the case of environmental pollution, therefore, exposure can occur in a number of different ways, i.e. by inhalation, ingestion or dermal absorption, and may involve a wide range of different organs. External exposure refers to the quantity of the pollutant at the interface between the recipient and the environment. The amount of any given pollutant that is absorbed is often termed the "absorbed dose", and may be dependent on the duration and intensity of the exposure. The "target organ dose" refers specifically to the amount that reaches the human organ where the relevant effects can occur (Sexton *et al.*, 1995). Exposure may be assessed in a range of different ways. External exposure is often measured using some form of personal monitor (e.g. passive sampling tubes for air pollution) or by modelling techniques (e.g. based upon knowledge of concentrations in the ambient environment). Biomarkers are indicators of exposure, dose, effect or susceptibility given by evidence found in biological samples (Links *et al.*, 1995). Sources of exposure data are discussed further in Chapter 5.

Historic data on pollution levels are often particularly sparse. Significant uncertainties in exposure classification consequently tend to occur, and the existence of a measurable concentration of a pollutant, even when higher than recommended levels, is not always a sufficient basis to infer health effects. Moreover, exposure often occurs to a number of different pollutants, in combination, and thus environmental concentrations of one pollutant do not always give a good indication of potential health effects. Social and other factors may also distort or mask the association between exposure and health outcome. Sexton *et al.* (1992) make several recommendations regarding the collection of data on human exposures. These include the need for standardised procedures for collection, storage, analysis and reporting; the involvement of different sectors for the design and maintenance of these databases; the collection of data at relevant levels of resolution (i.e. micro-environments where people are actually exposed); and the development of valid predictive models of exposure.

Effects

Exposure to environmental hazards, in turn, leads to a wide range of health effects (E_2). These may vary in type, intensity and magnitude depending upon the type of hazard to which people have been exposed, the level of exposure and the number of people involved. For convenience, a simple spectrum of effects can often be recognised. The earliest, and least intense, effects are sub-clinical, merely involving some reduction in function or some loss of well-being. More intense effects may take the form of illness or morbidity. Under the most extreme conditions, the result is death.

Health effect can be acute (e.g. microbiological contamination of water related to infant diarrhoea) or chronic (e.g. low levels of arsenic contamination in water related to cancer). Some contaminants may have a rapid effect following exposure, whereas others may require accumulation in the target organ before an adverse health effect can be observed. In such cases there may be a significant time lag (or latency period) between exposure and health effects. Health outcomes observed at present may be due to exposures which occurred many years or even decades earlier, as is the case with certain cancers, with consequent uncertainty regarding the actual dose the individual affected may have received (Rose, 1991).

One approach for assessing the impact of specific environmental exposures on health is quantitative risk assessment. Given known exposures and knowledge of dose–response functions, it is possible to make reasonable estimates of the health burden of specific pollutants. Further elaboration of risk analysis methods is needed, however, in order to provide a better basis for indicator development, by providing inexpensive and rapid estimates of the health impact of specific environmental exposures at the aggregate level (Nurminen and Corvalán, 1997).

Actions

In the face of environmental problems and observed health effects, society may attempt to adopt and implement a range of actions (A). These may take many forms and be targeted at different points within the environment–health chain. In the short term, actions are often primarily remedial (e.g. the treatment of affected individuals). In the longer term, actions may be protective (e.g. by trying to change individual behaviour and lifestyle to prevent exposure). Alternatively, actions may be taken to reduce or control the hazards concerned (e.g. by limiting emissions of pollutants or introducing measures of flood control). Perhaps the most effective long-term actions, however, are those that are preventative in approach, i.e. aimed at eliminating or reducing the forces which drive the system.

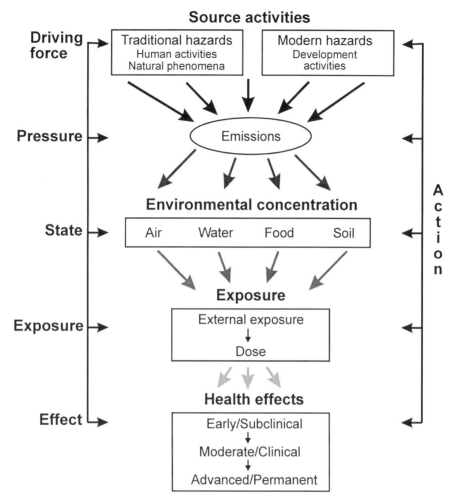

Figure 3.4 A simplified diagram of the environmental health hazard pathway and its link with the DPSEEA framework

3.4 Using the DPSEEA framework

In many situations, the causal pathway which has been described above can be quite complex; rather than a simple chain, it acts as a network of connections (Figure 3.4). For example, multiple effects may result from a single driving force (e.g. inadequate transport policies may lead to an increase in motor vehicle injuries, effects on the respiratory system, and noise disturbance). Equally, multiple causes may contribute to a single health effect (e.g. acute respiratory infections in children resulting from a combination of diverse driving forces, such as poverty, household policies, household energy policies, etc.) (WHO, 1997). In both these cases, the DPSEEA framework needs to be extended and adapted to include these multiplicity of pathways and links.

Against this background, Table 3.1 shows examples of indicators for one environmental health issue (occupational lead exposure). Note that the term "descriptive indicator" is used in Table 3.1 to describe indicators of driving forces, pressure, state, exposure and health effects, in the DPSEEA framework. Action indicators refer to actions at each level of the framework. A matrix of environmental health indicators, based on major and common driving forces, is also given in Table 3.2. This shows the range of indicators which might be developed for different environmental health issues, for each link in the DPSEEA framework.

As these example imply, an important question in developing any environmental health indicator is at what position within the DPSEEA framework it should be targeted. In terms of environmental epidemiology, the focus of attention is often the link between exposure and effect, for it is at this point that the environment is seen to have an impact on health. For this reason, it might be expected that most environmental health indicators are likely to be either exposure or effect indicators. To some extent this is true. In terms of health policy and management, however, it is often the earlier links in the DPSEEA framework which are of most interest. Many environmental health problems derive ultimately from relatively remote causal forces and events. Immediate sources of exposure thus represent little more than symptoms of the problem. Desertification, for example, is often a consequence of deeper-seated social and economic causes. Pollution, equally, is often a symptom of inadequacies in industrial technology and policy controls. If the aim is to identify the underlying cause of the problem, and to take effective action at source, it is therefore essential to have indicators that allow the effects on health to be traced back to their underlying sources and causes.

Indicators from higher up the DPSEEA framework also tend to provide a better early warning both of impending environmental problems and of the effects of intervention. Detectable changes in the state of the environment and in human health, for example, typically lag some way behind changes in source activity or emissions and in the case of some effects, such as cancers, often by many years. Most preventative action, similarly, occurs at or close to the source of the hazard (e.g. by controlling emissions at source or through hazard management).

A further reason for relying on indicators from higher up the DPSEEA framework is practical and is that of data availability. Typically, data become more difficult to acquire with each step down the chain. Thus, while there are normally abundant data on social and economic conditions and trends, much

Table 3.1   Examples of indicators within the DPSEEA framework: occupational lead exposure

Stage	Process	Descriptive indicator(s)	Action indicator(s)
Driving force	Type of development or human activities	Industrial/occupational use of lead Mining of lead	Technological innovation affecting use of lead Education about hazards of lead
Pressure	Source activities	Specific uses of lead Lead consumption (quantity produced and recycled)	Trends in lead use profile Trends in quantity of lead used Substitution for lead
	Emissions	Contamination of occupational and para-occupational environment	Availability and use of control technology
State	Environmental levels	Airborne lead concentrations Lead dust concentrations (work and home)	Trends in ambient air and dust monitoring
Exposure	Human exposure	Blood lead Blood ZPP Personal air sampling	Surveillance of blood lead and ZPP Trends in personal air monitoring Education about hazards of lead
	Dose	Blood lead Bone lead (research tool)	Trends in blood lead (e.g. government registries)
Effects	Early/subclinical	Deranged haem synthesis Non-specific CNS symptoms Abnormal nerve conduction velocity	Application of special surveys based in the workplace
	Moderate/clinical	Abdominal and constitutional symptoms Anaemia Decreased renal function	Routine medical surveillance (employment-based)
	Advanced/permanent	Renal failure Peripheral neuropathy Encephalopathy	Periodic analyses of major morbidity and mortality Clinical interventions

ZPP   Zinc protoporphyrin
CNS   Central nervous system

Table 3.2 Environmental health indicator matrix (illustrative example)

Driving force	Pressure(s)	State	Exposure(s)	Effect(s)	Action(s)
Population changes and social conditions	Social, economic, and demographic characteristics	Birth rate Age distribution Income distribution	Proportion of population living in poverty Proportion of population in vulnerable age groups (in association with other exposures)	Mortality, morbidity and disability (in association with other driving forces)	Education (particulary female) Health care Birth control initiatives Income distribution Equity policies
Human settlements and urbanisation	Urbanisation and urban migration Housing	Overcrowding Garbage disposal Noise levels Indoor pollution: – chemical – physical – biological	Proportion of population living in disadvantaged areas Proportion of time spent indoors Proportion of population living in affected housing	Road accidents Crime rate Infectious diseases Mental health Neurobehavioural disorders Cancer Respiratory conditions	Service provision Health facilities Facilitate growth of smaller urban centres Improved housing
Water requirements	Quantity: – inherent scarcity – increased consumption Quality: – natural – pollution (sewage, industrial effluent, urban run off and agricultural run off)	Water supply and sanitation: – formal access – private systems (e.g. wells) – informal market Industrial use Agricultural irrigation	Proportion of population without access to sanitation Proportion of population with insufficient water Proportion of population buying water from vendors	Morbidity and mortality resulting from: – water-borne diseases (e.g. cholera) – water-washed diseases (e.g. trachoma) – water-based diseases (e.g. schistosomiasis) – water-related diseases (e.g. malaria) – water-dispersed diseases (e.g. legionella)	Water conservation measures Use of urban wastewater for irrigation Increase access to safe water/hygienic sanitation Pollution control legislation Community education

Table 3.2 Continued

Driving force	Pressure(s)	State	Exposure(s)	Effect(s)	Action(s)
Food and agriculture needs	Food production and diet Amount produced	Calories per person Extent of land degradation Availability of water	Proportion of children with lower than acceptable calorie intake	Malnutrition Lower rate of growth in children Lowered immunity (Risks mostly in developing countries, and particularly for children)	Improved access and distribution Health education
	Microbiological contamination	Presence of microorganisms (measurements)	Consumption of contaminated food	Diarrhoea, typhoid fever, cholera, shigella etc. (Risk to the general population)	Access to clean water Improved personal hygiene, sanitation and hygenic food production (e.g. pasteurisation and irradiation)
	Toxic agents Type and amounts of chemicals used	Chemical additives Heavy metal releases in the environment Pesticides Agricultural chemicals and organic wastes contaminating water supply	Population living in affected areas Use (or lack of) of protective equipment for workers	Accidental poisoning Suicides (Risk particularly to workers and population in developing countries)	Legislation and supervision Improved labelling Use of protective clothing and equipment
Energy demand	Use of fossil fuels for transport, industry and home use (type and amount used)	Concentration of air pollutants (e.g. SO_2, PM_{10}, CO, NO_x, ozone, lead, cadmium mercury, arsenic)	Proportion of urban dwellers Proportion of population living in areas where these pollutants exceed recommended levels	Respiratory conditions, carcinogenic effects and other pollutant-specific morbidity/mortality effects (Risk to urban population)	Abatement expenditure Legislation for transport and industry Increased research into alternative power sources (e.g. solar and wind)

Table 3.2 Continued

Driving force	Pressure(s)	State	Exposure(s)	Effect(s)	Action(s)
Energy demand	Use of biomass fuel for cooking and heating (type and amount)	Concentration of indoor air pollutants (e.g. SO_2, PM_{10}, CO, NO_X, hydrocarbons, aldehydes, cresol, acenaphthylene, benzene, phenol, toluene, polyaromatic hydrocarbons)	Proportion of time spent indoors and in cooking areas	Respiratory conditions, CO poisoning and risk of respiratory cancer Accidental burns (Risks to women and children in both urban and rural settings in developing countries)	Improved access to improved stove designs Use of processed biomass fuels Use of fossil fuels (gas)
	Use of nuclear energy (amount of radioactive material used)	Number and state of facilities Radiation levels	Personal monitoring (workers) Population living in surrounding areas	Leukaemia and other cancers	Safety measures in place
Industry development	Workplace (characteristics, type of industry, type and amount of hazardous materials used)	Workplace exposure levels (e.g. asbestos, silica dust, organic solvents, lead, mercury, cadmium, manganese, arsenic nickel, aromatic amines, benzene, and noise)	Monitoring exposures in the workplace, in work-specific areas and in individual workers	Occupational diseases and accidents	Emission control measures Chemical safety legislation Epidemiologic studies Improved labelling Improved supervision
	Accidental releases (quantified emissions)	Short-term, high concentration of toxic substances (in air and water)	Environmental measures in populated areas	Several, including poisoning and cancer risk	Disaster prevention/ preparedness measures Environmental health impact assessment

Table 3.2 Continued

Driving force	Pressure(s)	State	Exposure(s)	Effect(s)	Action(s)
Industry development	Toxic chemicals and hazardous waste disposal (quantified)	Nature and amounts of hazardous materials in the environment (measured)	Population living around hazardous waste disposal sites	Several potential health effects (pollutant-specific)	Legislation for safe disposal methods Supervision (e.g. against illegal dumping)
Global limits	Release of CFCs and other ozone-damaging chemicals	Stratospheric ozone depletion Solar ultraviolet radiation at ground level	Proportion of time spent outdoors in specific locations Use of (or lack of) protection	Skin cancers Ocular cataracts Immunosuppression	Legislation (Montreal Protocol)
	Release of "greenhouse gases"	Climate change: – temperature and precipitation change – increased climate variability – sea level rise	Population living in affected areas	Heat-related illness and mortality Redistribution and re-emergence of vector- and water-borne diseases New and re-emerging infections Large-scale negative effects on nutrition	Research Monitoring Legislation (Framework Convention on Climate Change)

CFCs   Chlorofluorocarbons

less is known about the actual pressures on the environment, less still about environmental conditions and little about actual exposures. As a consequence, proxy indicators of exposure commonly have to be used that are derived from higher up the DPSEEA framework (Checkoway *et al.*, 1989).

The use of indicators from higher up the exposure chain, whether in their own right or as proxies, is not without its dangers. As noted earlier, to be effective any environmental health indicator must be based on a clear and firm relationship between the environmental hazard and the health effect. Unfortunately, the further removed the indicator is from the health effect, the weaker this link is liable to be. Each link in the chain is itself dynamic and uncertain; each step is subject to a wide range of influences and controls. The extent to which the driving forces are translated into active pressures on the environment, for example, depends upon the policy context, social attitudes and the pre-existing economic infrastructure of the area concerned. Whether these pressures cause detectable changes in the environment depends upon the ability of the environment to absorb and damp down the changes involved. Whether the environmental hazards, in turn, lead to health effects is determined by all the factors that control exposure and human susceptibility to its effects. It depends, therefore, on the form, duration, intensity and timing of exposure; on the social, economic and prior health status of the individuals concerned; and on the quality and accessibility of the health system. Equally, there is no certainty that action will be taken in response to the existence of environmental health problems. This depends not only on adequate recognition of the problems concerned, but also on political will, economic and technological capability and public acceptance of the actions involved. As a consequence, indicators from higher up the framework must be used and interpreted with care.

3.5   Conclusions

Population growth, technological and economic development, changing lifestyles and social attitudes, natural processes of change in the physical environment and the long-term impacts of past human interventions are all contributing to increasing problems of environmental health. To address these problems effectively, decision-makers require better information. This information needs to be reliable, consistent, targeted at the issues of real concern, available quickly, and available in an understandable and usable form.

Environmental health indicators provide one means of providing this information. In recent years, much progress has been made in developing indicators in a wide range of relevant fields and for many different applications. Progress in developing environmental health indicators has so far been slower, partly due to lack of consensus about the key issues that need to be addressed. There is, however, a growing need for environmental health indicators, both at the national and international level to inform broad-scale policy, and at the local scale in support of community- and city-level actions to improve and safeguard health.

Developing useful and effective indicators is a challenging task. Different users will have different expectations of the indicators they use, and a wide range of often competing criteria have to be met. The DPSEEA framework provides a useful means of rationalising the process of indicator design and construction. The next chapter considers the more technical issues involved in trying to apply these principles to indicator development in the area of environmental health.

3.6   References

Adriaanse, A. 1993 *Environmental Policy Performance Indicators. A Study of the Development of Indicators for Environmental Policy in the Netherlands.* Directorate-General for Environmental Protection, The Hague.

Bongaarts, J. 1994 Can the growing human population feed itself? *Scientific American*, March 1994, 18–24.

Briggs, D.J. 1995 Environmental statistics for environmental policy: genealogy and data quality. *Journal of Environmental Management*, **44**, 39–54.

Canadian Journal of Public Health 1991 Editorial — The demographic trap and sustainable health. *Canadian Journal of Public Health*, **82**, 3–4.

Carstairs, V. and Morris, R. 1989 Deprivation: explaining difference in mortality between Scotland and England and Wales. *British Medical Journal,* **299**, 886–9.

Checkoway, H., Pearce, N. and Crawford-Brown, D.J. 1989 *Research Methods in Occupational Epidemiology.* Oxford University Press, Oxford.

Corvalán, C., Kjellstrom, T. and Smith, K. 1999 Health, environment and sustainable development: Identifying links and indicators to promote action. *Epidemiology*, **10**, 656–60.

Environmental Research 1993 Environmental health agenda for the 1990s. Summary of workshop, October 1–6, 1991, Santa Fe, New Mexico. *Environmental Research*, **63**, 1–15.

Environment Canada 1991 *A Report on Canada's Progress Towards a National Set of Environmental Indicators.* SOE Report No. 91–1, Ministry of Environment, Ottawa.

Environmental Health Commission 1997 *Agendas for Change.* Chadwick House Group, London.

Gosselin, P., Bélanger, D., Bibeault, J.F. and Webster, A. 1993 Indicators for a sustainable society. *Canadian Journal of Public Health*, **84**, 197–200.

Gouzee, N., Mazijn, B. and Billharz, S. 1995 *Indicators of Sustainable Development for Decision-Making.* Federal Planning Office of Belgium, Brussels.

Harpham, T. and Blue, I. 1997 Linking health policy and social policy in urban settings: the new development agenda. *Transactions of the Royal Society of Tropical Medicine and Hygiene*, **91**, 497–8.

Hertz-Picciotto, I. 1996 Comment: Toward a coordinated system for the surveillance of environmental health hazards. *American Journal of Public Health*, **86**, 638–41.

House, J.S, Landis, K.R. and Umberson, D. 1988 Social relationships and health. *Science*, **241**, 540–5.

Jarman, B. 1983 Identification of underprivileged areas. *British Medical Journal*, **286**, 1705–9.

Kreisel, W.E. 1984 Representation of the environmental quality profile of a metropolitan area. *Environmental Monitoring and Assessment* **4**, 15–33.

Last, J.M. [Ed.] 1995 *A Dictionary of Epidemiology*. Oxford University Press, Oxford.

Links, J.M., Kensler, T.W. and Groopman, J.D. 1995 Biomarkers and mechanistic approaches in environmental epidemiology. *Annual Review of Public Health*, **16**, 83–103.

Lioy, P.J. 1990 Assessing total human exposure to contaminants. *Environmental Science and Technology,* **24**, 938–45.

Litsios, S. 1994 Sustainable development is healthy development. *World Health Forum*, **15**, 193–5.

Local Government Management Board 1994 *Sustainability Indicators - Guidance to Pilot Authorities*. Touche Ross, London.

McMichael, A J. 1993 *Planetary Overload — Global Environmental Change and the Health of the Human Species*. Cambridge University Press, Cambridge.

Mullan, R.J. and Murthy, L.I. 1991 Occupational sentinel health events: an up-dated list for physician recognition and public health surveillance. *American Journal of Industrial Medicine* **19**, 775–99.

National Academy of Sciences 1991 *Human Exposure Assessment for Airborne Pollutants. Advances and Opportunities*. National Academy of Sciences, National Academy Press, Washington, D.C.

Nurminen, M. and Corvalán, C. 1997 Methodological issues in using epidemiological studies for health risk analysis. In: C. Corvalán, M. Nurminen and H. Pastides [Eds] *Linkage Methods for Environment and Health Analysis — Technical Guidelines*. World Health Organization, Geneva.

OECD 1993 *OECD Core Set of Indicators for Environmental Performance Reviews*. Environmental Monograph No 83, Organisation for Economic Co-operation and Development, Paris.

OECD 1997 *Better Understanding Our Cities — The Role of Urban Indicators*. Organisation for Economic Co-operation and Development, Paris.

Ostro, B. 1994 Estimating the health effects of air pollution. A method with an application to Jakarta. Policy Research Working Paper No. 1301. The World Bank, Washington D.C.

Pastides, H. 1995 An epidemiological perspective on environmental health indicators. *World Health Statistics Quarterly*, **48**, 140–3.

Pirkle, J.L., Needham, L.L. and Sexton, K. 1995 Improving exposure assessment by monitoring human tissues for toxic chemicals. *Journal of Exposure Assessment and Environmental Epidemiology*, **5**, 405–24.

Projecto Estado de la Nacion 1995 *Estado de la Nacion en desarrollo humano sostenible*. Projecto Estado de la Nacion, Costa Rica.

Rapport, D.J. 1992 Evolution of indicators of ecosystem health. In: D.H. MacKenzie, D.E. Hyatt and V.J. McDonald [Eds] *Ecological Indicators, Volume 1*. Elsevier Applied Science, London, 121–33.

Rose, G. 1991 Environmental health: problems and prospects. *Journal of the Royal College of Physicians of London,* **25**, 48–52.

Rothwell, C.J., Hamilton, C.N. and Leaverton, P.E. 1991 Identification of sentinel health events as indicators of environmental contamination. *Environmental Health Perspectives,* **94**, 261–3.

Sexton, K., Callahan, M.A. and Bryan, E.F. 1995 Estimating exposure and dose to characterise health risks: the role of human tissue monitoring in exposure assessment. *Environmental Health Perspectives*, **103**(S3), 13–29.

Sexton, K., Selevan, S.G., Wagener, D.K. and Lybarger, J.A. 1992 Estimating human exposures to environmental pollutants: availability and utility of existing databases. *Archives of Environmental Health*, **47**, 398–407.

Sexton, K., Wagener, D.K., Selevan, S.G., Miller, T.O. and Lybarger, J.A. 1994 An inventory of human exposure related data bases. *Journal of Exposure Analysis and Environmental Epidemiology*, **4**, 95–109.

Stephens, C. 1995 The urban environment, poverty and health in developing countries. *Health Policy and Planning*, **10**, 109–121.

Sustainable Seattle 1993 *The Sustainable Seattle Indicators of Sustainable Community*. Sustainable Seattle, Seattle.

SEPA (Swedish Environmental Protection Agency) 1993 *Environment and Public Health. An Epidemiological Research Programme*. Swedish Environmental Protection Agency, Solna.

Thacker, S.B., Stroup, D.F., Parrish, R.G. and Anderson, H.A. 1996 Surveillance in environmental public health: issues, systems and sources. *American Journal of Public Health*, **86**, 633–8.

UNDP (United Nations Development Programme) 1990 *Human Development Report*. Oxford University Press, New York.

UNDP (United Nations Development Programme) 1997 *Human Development Report*. Oxford University Press, New York.

UNEP 1993 *United Nations Environment Programme Environmental Data Report 1993–1994.* Blackwell Publishers, Oxford.

UNEP/RIVM 1994 An overview of environmental indicators: state of the art and perspectives. UNEP/EATR.94-01, United Nations Environment Programme, Nairobi; RIVM/402001001, National Institute of Public Health and Environment, Bilthoven.

UNEP/WHO 1993 GEMS/Air — Global environmental monitoring system: a global programme for urban air quality monitoring and assessment. Doc. No. WHO/PEP 93.7, World Health Organization, Geneva; UNEP/GEMS/93.A.I. United Nations Environment Programme, Nairobi.

United Nations 1993 *Agenda 21: Programme of Action for Sustainable Development.* United Nations, New York.

United Nations 1996 *Indicators of Sustainable Development — Framework and Methodologies.* United Nations, New York.

US EPA 1994 *A Conceptual Framework to Support the Development and Use of Environmental Information.* United States Environmental Protection Agency, Triangle Park, North Carolina, USA.

Warford, J.J. 1995 Environment, health, and sustainable development: the role of economic instruments and policies. *Bulletin of the World Health Organization,* **73**, 387–95.

WHO 1987 *Air Quality Guidelines for Europe.* WHO Regional Publications, European Series No. 23., World Health Organization, Copenhagen.

WHO 1990 Joint UNEP/FAO/WHO food contamination monitoring programme. Doc. No. WHO/EHE/FOS/90.2, World Health Organization, Geneva.

WHO 1991 GEMS/Water 1990–2000: The challenge ahead. Doc. No. WHO/PEP/91.2, World Health Organization, Geneva.

WHO 1993a *Implementation of Strategies for Health for All by the Year 2000. Third Monitoring of Progress. Common framework.* World Health Organization, Geneva.

WHO 1993b *Implementation of the Global Strategy for Health for All by the Year 2000, Second Evaluation — Eighth Report on the World Health Situation.* World Health Organization, Geneva.

WHO 1997 *Health and Environment in Sustainable Development — Five Years After the Earth Summit.* World Health Organization, Geneva.

Wills, J. and Briggs, D. 1995 Developing indicators for environment and health. *World Health Statistics Quarterly,* **48**, 155–63.

Winkelstein, W. 1992 Determinants of worldwide health. *American Journal of Public Health,* **82**, 931–2.

World Bank 1992 *World Development Report 1992: Development and the Environment.* Oxford University Press, Oxford.

World Bank 1994 *Monitoring Environmental Progress: a Report of Work in Progress.* The World Bank, Washington.

World Resources Institute 1995 *Environmental Indicators: A Systematic Approach to Measuring and Reporting on Environmental Policy Performance in the Context of Sustainable Development.* World Resources Institute, New York.

World Wide Fund for Nature (WWF) and The New Economics Foundation 1994 *Indicators for Sustainable Development.* World Wide Fund For Nature, London.

Chapter 4[*]

METHODS FOR BUILDING ENVIRONMENTAL HEALTH INDICATORS

4.1   The challenge of environmental health indicators

As the previous chapter has illustrated, the development of reliable and effective environmental health indicators is not a trivial task. To be effective, they must meet a range of criteria (as outlined in Box 3.1). They must be matched to their purpose: i.e. they must address the problem of concern, at the appropriate point in the environment–health chain, and at appropriate geographical and temporal scales and resolution. Both the data and the computational methods and models needed to construct them must be available at an acceptable cost. They must be expressed and presented in an easily understandable and usable form, and must be scientifically valid and testable. Moreover, if the results of indicators are to be more widely applicable, if the indicators themselves are to be accepted by the many stakeholders concerned (e.g. scientists, politicians, the public), and if lessons are to learned from the collective experience in developing and using indicators, it is important that all these issues of design are carefully documented and open to scrutiny.

The above requirements have significant implications for the way in which indicators are designed and constructed. Many of the criteria are also to some extent mutually incompatible; that is one reason why indicators are difficult to design. The ultimate need for cost-effectiveness, for example, often means that indicators must be developed on the basis of data that already exist or which (if newly collected) can also be used for other purposes. Unfortunately, many of the data that do exist have been collected for specific purposes, and are therefore not ideal for other applications. The need for clarity and ease of understanding also implies that indicators must often condense large volumes of data into a brief overview, and reduce the complexities of the world to a simple and unambiguous message. The need for scientific validity, on the other hand, requires that this process of *précis* must not go too far. Indicators must simplify without distorting the underlying truth, or losing the vital connections and interdependencies which

---

[*]   *This chapter was prepared by D. Briggs*

govern the real world. At the same time, if indicators are to be sensitive to change, they need to be based on accurate, high resolution and consistent data. Achieving this, whilst also maintaining simplicity, is itself a challenge. To do so whilst also ensuring that the indicators can make use of the limited, and often varied, data that are usually available is even more difficult. To achieve all this cost-effectively is difficult indeed.

Careful indicator design thus holds the key to success. As with any piece of engineering, it is also important to appreciate that design represents an attempt to balance two sets of requirements: the need of the user and the constraints or limitations of the materials and technology (in this case the data and models) which are available. Good indicator design is therefore not a case of simply defining an indicator which reflects the users concerns — it must be practicable to construct and use. Nor is it a matter merely of repro-ducing whatever data happens to be available — the data need to be manipulated and customised as far as possible to meet the users' needs. It is probably fair to argue that, in the past, inadequate attention to indicator design has meant that many indicators have been somewhat ineffective, and many have not got beyond the proposal stage. This chapter explores some of the issues involved in designing good and usable environmental health indicators in the face of these challenges. It considers the types of indicators that might be constructed, outlines the steps in indicator design and construc-tion, and presents examples of environmental health indicators in the form of "indicator profiles".

4.2   Types of indicator

Environmental health indicators may take many forms. The previous chapter has already drawn a distinction between exposure-based and effect-based indicators. The former provide measures (albeit indirect) of exposure to some risk factor, from which it is possible to deduce a potential health effect; the latter provide measures of some health outcome, for which it is possible to infer an environmental cause. Environmental health indicators may also be designed, however:

- To detect temporal trends or spatial patterns.
- As simple or composite indicators.
- At the local, national or international scale.
- For the purpose of policy/management, epidemiological research or awareness raising.

4.2.1   Temporal versus spatial indicators

One of the most important distinctions in the case of indicators is that between temporal and spatial indicators. Each has a somewhat different purpose and tends to be used in different ways.

Temporal indicators are designed to describe and measure changes over time. They are therefore often used as a means of detecting time-based trends in environmental health (e.g. to see whether conditions are getting better or worse), to predict change and give an early warning of new or emerging problems, to monitor the effects of changing circumstances (e.g. driving forces) on environmental health, to monitor the effects of policy or management interventions, or to check on compliance with environmental legislation. They may also be used to examine relationships between environment and health, using time-series methods (see Chapter 6).

Spatial indicators, in contrast, are intended to identify and describe geographic variations. They may be used, for example, to help identify systematic spatial patterns in environmental health (spatial trends), to detect "hotspots" or "clusters", to compare different areas in terms of their environmental health status, or to compare the effects of different policy and management strategies. As part of epidemiological research studies, they may also be used as a basis for investigating associations between environment and health, using ecological methods (Chapter 6).

These two types of indicator require different designs, pose different data needs and are likely to be expressed in different ways. Indicators of air quality (e.g. atmospheric concentrations of fine particulates) provide a useful example. Designed as a temporal indicator, these rely on repeated measurements taken at a sample of sites, over time. Measurements will need to be taken frequently enough to show the changes of interest, and for many purposes may need more-or-less continuous monitoring. The spatial distribution of the sampling or measurement points needs to be representative, but they do not necessarily need to provide intensive or complete geographical coverage. Ideally, the spatial distribution of these data will need to be fixed, so that changes due to the geographic location of the measurement sites do not cloud the picture (the user needs to be confident that all the variation in the indicator relates to real changes over time and not to changes in the monitoring network). Detailed information on the spatial distribution or location of the monitoring sites may not be essential. Differences in the methods used to measure pollution levels at different sites may also be less important, because data will only be compared with other data from that site (or group of sites). In general, temporal indicators will be good for examining acute health effects; they will be less useful for measuring or describing chronic effects. The indicator is often best expressed as a line-graph or histogram.

An air quality indicator designed to examine geographic patterns and identify "hotspots", however, will have very different requirements. In this case, the spatial distribution of the data becomes crucial. The monitoring sites will need to be geographically intensive, and cover the whole area of interest. The sampling density will depend on how much the pollutant varies spatially, and the smallest size of area which needs to be detected. Often, the distribution of sites will need to be stratified in some way, so that sampling is more intensive in those areas where spatial variation is greatest. It will be essential that all sites use the same monitoring techniques, otherwise spurious patterns will occur in the data. Accurate data on the geographic location of the sites is also essential. Temporal characteristics of the data, on the other hand, are less restrictive. Continuous monitoring may not be necessary; instead, the interest may simply be in estimating the long-term average (e.g. annual mean concentration) or the level of pollution within a specified period. The timing of the monitoring will need to be standardised, however, so that measurements are representative of the period of interest. Spatial indicators of this sort thus tend to be most useful for examining chronic health effects; unless monitoring is repeated frequently, they are unlikely to provide useful measures of acute effects. The indicator is often best presented as a map, or series of maps, which can show the spatial variations in conditions.

This distinction between spatial and temporal indicators is not wholly valid. Some indicators might be intended to serve both purposes: to show spatial variations in temporal trends, or to show changes in the geographic pattern over time. Developing and constructing spatio-temporal indicators of this type is, however, difficult because the data demands become severe — data need to be both spatially and temporally comparable, intensive and accurate. In many cases, data that meet these standards are not available.

4.2.2   Simple versus composite indicators

Indicators represent an attempt to simplify the complexity of reality into an easily interpretable measure. In order to describe reality, however, a large number of different indicators could be needed, relating for example to the many different hazards and health outcomes of interest and to the different points in the DPSEEA chain. In using specific indicators of this type, therefore, there is the possibility that decision-makers will be confronted with a bewildering range of information, much of it apparently contradictory in the message it gives.

In the light of this problem, there have been many attempts in recent years to develop more synoptic or composite indicators, which condense a wide range of information on different (but related) phenomena into a single measure or index. An often quoted example of this is the Retail Prices Index, which is used to show trends in inflation based upon a "basket" of goods. A composite indicator of human development has similarly been developed by UNDP (1990). Other examples are the various indicators of deprivation which have been widely used in social sciences and epidemiology (e.g. Townsend, 1987; Jarman, 1984; Carstairs and Morris, 1991) and the composite indicators of environmental quality developed by Inhaber (1976) and Hope and Parker (1991). Composite indicators of this type are already used in measuring land suitability (e.g. FAO, 1976). Similarly, many countries use composite indicators of stream water quality, based either on a range of chemical parameters or on their biological status (Newman, 1992).

A similar case can be made for the development and use of composite environmental health indicators. Because of the need for an explicit linkage with health, it is unlikely that the sorts of general indicator of overall environmental quality which have so far been proposed are of much use. On the other hand, it may be useful in some cases to construct composite indicators either of total exposure to a specific hazard (i.e. covering all media and exposure pathways) or of groups of hazards. Thus, instead of producing separate indicators for exposure to each air pollutant, it might be possible to derive a composite indicator of exposure, including all air pollutants of interest. Ostensibly, indicators of this type have a number of benefits. By reducing the volume of information, for example, they facilitate the decision-maker's task. Equally, by taking account of the various pollutants to which people may be simultaneously exposed, they offer the scope to allow for additive and synergistic effects. Composite health indicators are also possible, either by combining data for a number of different diseases (e.g. using a wider range of ICD codes (WHO, 1992)), or by combining data on morbidity and mortality within a single index.

Nevertheless, composite indicators also have many dangers and disadvantages. One problem is that such indicators require more data and the indicator is thus more than ever susceptible to gaps or weaknesses in data availability. More importantly, the results of the indicator depend to a great extent upon how it is constructed, what variables are used, and how these are weighted and combined. Where the different components of the indicator are measured in the same units, it is theoretically feasible to combine them by simple addition or averaging. For example, the total pollutant concentration in the air can be calculated by summing the concentrations (in parts per million or micrograms per cubic metre) of all the pollutant species of interest. Such a process does not necessarily make sense, however, because it assumes that all pollutants are of equal importance. Composite indicators may also attempt to bring together different components which are measured on different scales, so that simple arithmetic manipulation is not feasible. In these circumstances, some form of model or combination procedure needs to be developed by which to construct the indicator. Commonly, this involves some form of weighted aggregation, the weights being derived either empirically (e.g. from regression analyses), from first principles or by expert judgement. Where the indicator is intended to provide an index of health risk, these weights might be chosen to reflect the known harmfulness of each hazard (e.g. toxicity of each pollutant), although even this causes problems because the different

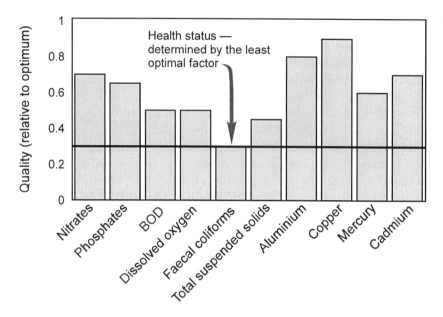

Figure 4.1 The principle of limiting factors: the example of water quality and health

pollutants may have different health effects. Complex interactions may also occur between the various pollutants, so that the overall effect on health cannot simply be conceived as the sum of the various parts. In the case of effect-based indicators, disability adjusted life years (DALYs) (World Bank, 1993; Murray and Lopez, 1996) may be used to provide a common measurement scale by which to combine data (both on mortality and morbidity) for different diseases (see Chapter 8). This is only valid, however, where a consensus exists about the relative severity of the different health effects.

Another possible approach is one based on the principle of limiting. This assumes that the condition of interest is defined by the state of the worst (or least optimal) factor. Figure 4.1 shows a theoretical example. In this case, the indicator "percentage of people with access to safe drinking water" assumes that all pollutants of potential health concern are below specified limits; exceedance of these limits by any one pollutant would render the water "unsafe". Equally, an indicator such as "number of days of clean air" might be conceived to give a general measure of levels of urban air pollution. Again, if any of the pollutants of concern exceeded recommended limits, the air would be classified as "not clean".

The choice of model for compiling composite indicators of this type is clearly crucial. Unless an accepted model exists by which to convert the various components to a common measurement scale (e.g. to comparable measures of risk), the construction of such indicators is clearly likely to be somewhat arbitrary and open to challenge. It may also be difficult to test or verify composite indicators, because they do not relate to specific, measurable conditions. For the same reasons, it is difficult to establish clear standards and guidelines for composite indicators of this type and as a consequence interpretation of composite indicators can be a problem.

4.2.3   Local, national and international indicators

Environmental health indicators can be used at a wide range of geographic scales: from the level of an individual community, to that of a city, to a wider administrative or geographic region, to a country or even the whole world. The scale of application will in each case have considerable implications for the way in which the indicator is designed and constructed.

Amongst the most important issues relating to the geographic scale of aggregation are the spatial extent and resolution of the data needed to construct the indicator. At the local level, spatial resolution is often the major consideration: can data be obtained at a fine enough resolution to show accurately conditions within the area of interest? This is often problematic. Several of the HEADLAMP field studies, reviewed in Chapter 8 for example, showed that important health-related data were only available at a relatively coarse level of aggregation (e.g. regional or national), meaning that locally specific indicators could not easily be developed. To some extent this problem can be overcome by using GIS techniques to disaggregate data (see Chapter 7). However, disaggregation must always be carried out with care, because it invariably involves making some assumptions about the geographic distribution of conditions at the local scale.

At the national and even more so at the international level, different data issues tend to arise. In these cases, the major problem is often to obtain sufficient data, in a consistent and comparable form, across the whole area of interest. Differences in measurement techniques, data definitions, sampling regime and all the other factors that affect data quality and comparability can pose difficulties at this scale. In the case of health data, for example, there may be important, although hidden, discrepancies in rates of referral, diagnosis of diseases, and the effectiveness of reporting systems between different countries, which render international comparisons of health data potentially misleading, notwithstanding the efforts of WHO. Environmental monitoring and survey techniques vary enormously between countries, and often even within countries (Briggs, 1995). Some international networks do exist which have endeavoured to establish standardisation (e.g. the Global Resource Information Database (GRID) network) (GRID, 1999), but often these are sparse and potentially biased in their distribution, so they cannot be relied upon to provide representative data across the world (or even the participating countries).

4.2.4   Policy, epidemiological and awareness-raising indicators

As has already been noted, environmental health indicators may serve a wide range of purposes. They may be used to support and inform management and policy, as part of epidemiological investigations of associations between environment and health, and as a means of raising awareness about specific issues or interests. Again, these different uses imply the need for differences in indicator design: the criteria listed in Box 3.1 each assume different levels of importance, and indicators may be targeted at different points in the DPSEEA chain.

In the case of policy and management, for example, indicators will usually be required that clearly address issues falling within the remit of those concerned, and which are open to influence and control. Action indicators will be especially important, because one of the main needs is to monitor and assess the policy actions taken and to evaluate their effects. Because effective management and policy often requires preventative action, there is likely to be a preference in many cases for indicators relating to the upper links in the chain, namely the driving forces and pressures. Because policy-makers also need to know if these interventions have been effective in reducing health risks, health effect indicators will also be useful. Measures of exposure, however, may be seen as less vital.

Epidemiological research will often have a different focus. Typically, attention is targeted at the specific relationship between environmental exposure and health outcome. The need is thus for direct indicators of exposure and health effect. Indicators from higher up the chain, such as measures of state or pressure, will often be used only as proxies where direct measures of exposure are not available. Scientific validity and accuracy are paramount; issues such as the complexity or "resonance" of the indicator are far less significant.

Different considerations tend to apply in the case of indicators designed to promote public awareness. Here, the main need is often for resonance (i.e. the indicators need to be interesting and acceptable to the community concerned). Indicators will need to relate directly to issues that are of concern to the community, and will need to be expressed in ways which they understand. Often, this will mean that some degree of complexity and rigour may need to be sacrificed to make the message bold and clear.

4.3   Steps in indicator development

A large number of questions clearly have to be faced in designing and using indicators. The details of these questions vary depending upon the particular character of the indicator and its intended use. As a result, there is no rigid and universal process of indicator construction. Figure 4.2, however, summarises the steps commonly involved. As this figure indicates, the main steps are as follows:

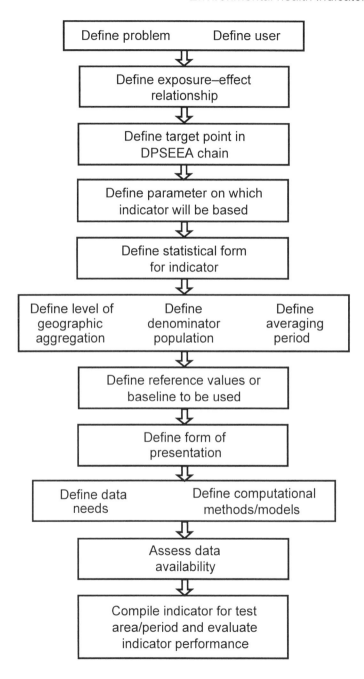

Figure 4.2 Steps in the construction of environmental health indicators

1. Specification of the problem to be addressed (i.e. the use of the indicator) and the user(s) concerned. The purpose might be defined in various ways, depending upon the interests of the user; for example in terms of a specific environmental hazard (e.g. ionising radiation), a specific health outcome (e.g. childhood leukaemia), a specific policy or action (e.g. food hygiene legislation) or an underlying driving force (e.g. population growth).

2. Specification of the environment–health relationship on which the indicator will be based. This is essential if a valid environmental health indicator is to be identified. The relationship may, however, be expressed in more or less quantitative terms (e.g. as an explicit exposure–effect relationship) or as a general tendency (e.g. for poor sanitation to lead to higher rates of infection).

3. Specification of the point in the DPSEEA framework at which the indicator will be targeted. This will depend upon the particular interest and responsibilities of the user, but will also be influenced by the availability of relevant data and computational methods.

4. Specification of the parameter on which the indicator will be based — i.e. the particular measure of environment or health which will be used (e.g. atmospheric NO_2 concentration, cough and wheeze).

5. Specification of the statistical form of the indicator. This step involves a number of considerations. Indicators can be presented in a variety of statistical forms, such as simple frequencies or magnitudes (e.g. number of deaths), as rates (e.g. emission rates, mortality rates), as ratios (e.g. pollution level relative to the WHO guideline level, standardised mortality ratio), as measures of rate change (e.g. rate of population growth, rate of reduction in air pollution level), or in various more complex forms. The form chosen should reflect the purpose of the indicator.

6. Specification of the denominators and levels of aggregation required for the indicator (e.g. the level of geographic aggregation, denominator population, averaging period).

7. Specification of the baseline or reference data against which the indicator will be standardised. This will need to reflect the statistical form of the indicator and the level of geographic aggregation, etc.

8. Specification of the form in which the indicator will be presented (e.g. graphically, as a map, as a simple statistic).

9. Specification of the data needs and models or methods required to compute the indicator.

10. Assessment of data availability and quality in the light of the foregoing specifications. At this stage, if relevant data are unavailable, it may be necessary to reconsider the indicator design (e.g. by choosing a proxy or by using a different level of aggregation).

11. Computation and testing of the indicator for a pilot area. This is a crucial step in order to determine whether the indicator is sensitive to the variations in the conditions of interest, whether the computational methods are sufficiently robust and the data adequate, and whether the results of the indicator are interpretable.

For the sake of clarity, these are presented here as a simple sequence. In reality, however, they are normally interactive and reiterative in form. Many of the questions of indicator design are interdependent, and need to be considered simultaneously. Many aspects of indicator design ultimately have to be amended in response to practical issues, such as data availability. Until the indicator has been tested and used, it may not be certain that it operates effectively.

One aspect of this process needs to be emphasised: namely quality control. In that environmental health indicators contribute directly to decisions about human welfare and health, they inevitably carry a heavy burden of responsibility. Far-reaching and costly consequences can flow from their use. The validity of environmental health indicators is therefore of paramount concern. The construction and use of environmental health indicators thus need to include provisions for validation and quality control. These need to consider not only the way in which the indicator is designed but also the data, methods and models used in its application.

Processes of quality control for indicators are not especially well-developed. Ideally, however, the data sources used need to be checked (e.g. by examining the genealogy of the data and by cross-validating the data against independent sources). As far as possible, the indicators themselves should also be tested for inconsistencies. Trends and geographic distributions should be inspected carefully to identify significant discontinuities, and these should be investigated to ensure that they are not artefacts of the data sources or methods used. Comparisons should be made between indicators to check for unexpected departures from established relationships. The definition of indicators should be checked to ensure comparability. Where feasible, margins of error should be assessed so that the true patterns or trends can be separated from "noise" due to uncertainty in the indicators. The definitions, methods and data sources used in constructing environmental health indicators should always be fully documented, in order to facilitate these quality checks. This process of quality control is not restricted to the design and construction stages. It needs to be continued as long as the indicator is being used, both to ensure that changes in the data, real-world conditions or level of knowledge have not rendered it invalid, and to ensure that it is still providing useful information.

4.4   Towards a core set of environmental health indicators

In recent years, considerable effort has been devoted to developing core sets of environmental and sustainability indicators for policy support. It might therefore be expected that similar core sets of environmental health indicators could usefully be constructed. Establishment of a core set of indicators would certainly offer a number of advantages:

- They could save time and resources, by avoiding duplication of effort in researching and developing new indicators.
- They could provide a basis for comparison between different areas and over time.
- They could help to establish standards for indicator development which will improve the general quality of information available to the decision-maker.

In practice, the construction of a core set of environmental health indicators is a much more difficult task than may be supposed. By definition, indicators need to be use-specific, and therefore indicators developed for one application cannot readily be translated to another. Indicators tend to be driven by prior concern about a problem. In some areas of application, such as the environment and economy, a broad consensus often exists about what these key problems are. Core sets of indicators can thus be developed on this basis. In the area of environmental health, however, this consensus is less well established, and many of the problems may be relatively local in their extent. The definition of core environmental health indicators is therefore more difficult.

As emphasised earlier, environmental health indicators also need to be based upon known and definable associations between environment and health. These associations have often proved difficult to establish, except at a local level, due to the complexities of confounding and the problems in acquiring reliable environmental and health data at an appropriate spatial and temporal resolution. Many potential environmental health indicators are thus of limited use due to uncertainties in the environment–health linkages on which they are based.

For these reasons, no attempt here is made to present formal lists of core indicators. Instead, indicator profiles are presented for a selection of environmental health indicators, relating to a range of different environmental health issues (Table 4.1). These profiles are not intended to be definitive, instead they serve as a model which can be developed and customised according to need. Table 4.2 presents a key to these indicator profiles. As this shows, the profiles are designed to provide a range of information on

Table 4.1 Summary list of examples of environmental health indicators

Issue/theme or topic	Indicator	Example definition	DPSEEA
Socio-demographic context			
Poverty	Human poverty index	Human poverty index	Driving force
Population density	Population density	Population density	Driving force
Population growth	Rate of population growth	Annual net rate of population growth	Driving force
Age structure	Dependent population	Percentage of people aged less than 16 years or 65 years or more	Driving force
Urbanisation	Rate of urbanisation	Annual net rate of change in the proportion of people living in urban areas	Driving force
Infant mortality	Infant mortality rate	Annual death rate of infants under one year of age	Effect
Life expectancy	Life expectancy at birth	Number of years a newborn baby is expected to live, given the prevailing mortality rate	Effect
Air pollution			
Outdoor air pollution	Ambient concentrations of air pollutants in urban areas	Mean annual concentrations of SO_2, NO_2, O_3, CO, particulates (PM_{10}, $PM_{2.5}$, SPM) and lead in the outdoor air in urban areas	State
Indoor air pollution	Sources of indoor air pollution	Percentage of households using coal, wood or kerosene as the main source of heating and cooking fuel	Exposure
Respiratory illness	Childhood morbidity due to acute respiratory illness	Incidence of morbidity due to acute respiratory infections in children under five years of age	Effect
Respiratory illness	Childhood mortality due to acute respiratory illness	Annual mortality rate due to acute respiratory infections in children under five years of age	Effect
Air quality management	Capability for air quality management	Capability to implement air quality management	Action
Air quality management	Availability of lead-free petrol	Consumption of lead-free petrol as a percentage of total petrol consumption	Action

Continued

Table 4.1 Continued

Issue/theme or topic	Indicator	Example definition	DPSEEA
Sanitation			
Excreta disposal	Access to basic sanitation	Proportion of the population with access to adequate excreta disposal facilities	Exposure
Diarrhoea	Diarrhoea morbidity in children	Incidence of diarrhoea morbidity in children under five years of age	Effect
Diarrhoea	Diarrhoea mortality in children	Diarrhoea mortality rate in children under five years of age	Effect
Shelter			
Informal settlements	Population living in informal settlements	Percentage of the population living in informal settlements	Exposure
Unsafe housing	Percentage of the population living in unsafe housing	Percentage of the population living in unsafe, unhealthy or hazardous housing	Exposure
Home accidents	Accidents in the home	Incidence of accidents in the home	Effect
Urban planning	Urban planning and building regulations	Scope and extent of building regulations for housing	Action
Access to safe drinking water			
Water quality/supply	Access to safe and reliable supplies of drinking water	Percentage of the population with access to an adequate amount of safe drinking water in the dwelling or within a convenient distance from the dwelling	Exposure/ action
Water quality/supply	Connections to piped water supply	Percentage of households receiving piped water to the home	Exposure/ action
Diarrhoea	Diarrhoea morbidity in children	Incidence of diarrhoea morbidity in children under five years of age	Effect
Diarrhoea	Diarrhoea mortality in children	Diarrhoea mortality rate in children under five years of age	Effect

Continued

Table 4.1 Continued

Issue/theme or topic	Indicator	Example definition	DPSEEA
Access to safe drinking water (continued)			
Water-borne diseases	Incidence of outbreaks of water-borne diseases	Incidence of outbreaks of water-borne diseases	Effect
Water quality monitoring	Intensity of water quality monitoring	Number of valid measurements of water quality per thousand head of population per year	Action
Vector-borne disease			
Population at risk	Population at risk from vector-borne diseases	Number of people living in areas endemic for vector-borne diseases	Exposure
Vector-borne disease mortality	Mortality due to vector-borne diseases	Mortality rate due to vector-borne diseases	Effect
Vector control	Adequacy of vector control systems	Percentage of the at-risk population covered by effective vector control and remediation systems, by disease and programme type	Action
Waste management			
Waste collection	Municipal waste collection	Percentage of the population served by regular waste collection services	Action
Waste disposal	Municipal waste disposal	Mass of solid waste disposed of by municipal waste management services	Action
Waste management	Hazardous waste policies	Effectiveness of hazardous waste policies and regulations	Action
Hazardous/toxic substances			
Blood lead	Blood-lead level in children	Percentage of children with blood lead levels of more than 10 µg per 100 ml	Exposure
Chemical poisonings	Mortality due to poisoning	Mortality rate due to poisoning	Effect
Contaminated land	Contaminated land management	Scope and rigour of contaminated land management	Action

Continued

Table 4.1 Continued

Issue/theme or topic	Indicator	Example definition	DPSEEA
Food safety			
Food-borne diseases	Food-borne illness	Incidence of outbreaks of food-borne illness	Effect
Diarrhoea	Diarrhoea morbidity in children	Incidence of diarrhoea morbidity in children under five years of age	Effect
Diarrhoea	Diarrhoea mortality in children	Diarrhoea mortality rate in children under five years of age	Effect
Monitoring of food	Monitoring of chemical hazards in food	Proportion of potentially hazardous chemicals monitored in food	Action
Radiation			
Radiation exposure	Cumulative radiation dose	Percentage of the population receiving an effective radiation dose in excess of 5 mS a^{-1}	Exposure
UV exposure	UV light index	UV light index	Exposure
Non-occupational health risks			
Motor vehicle accidents	Mortality from motor vehicle accidents	Death rate due to road accidents	Effect
Non-occupational injury	Injuries to children	Incidence of physical injury to children less than five years of age	Effect
Poisoning	Incidence of poisonings of young children	Number of reported poisonings in children under five years of age per year	Effect
Occupational health risks			
Occupational hazards	Exposure to unsafe workplaces	Percentage of workers exposed to unsafe, unhealthy or hazardous working conditions	Exposure
Occupational morbidity	Morbidity due to occupational health hazards	Incidence of occupational injury	Effect
Occupational mortality	Mortality from occupational health hazards	Incidence of occupational mortality	Effect

Table 4.2   Key to indicator profiles

Brief title of indicator	Position in DPSEEA chain

Indicator profile

Issue	Specification of the environmental health issue(s) to which the indicator relates.
Rationale and role	Outline of the justification for the indicator and its potential use in relation to the issue(s) specified. Where appropriate, indicate the main user communities and the level of aggregation/geographic scale at which the indicator might be used.
Linkage with other indicators	Describe the relationship between this and other indicators relating to the issue(s) specified, listing all indicators and their position in the DPSEEA chain.
Alternative methods and definitions	Outline possible methods for defining and constructing the indicator. In particular, suggest how the indicator can be improved (where suitable data exist), or adjusted/simplified to cope with inadequacies in the available data. If appropriate, suggest proxy indicators.
Related indicator sets	List similar or related indicators, proposed or developed as part of other indicator sets (e.g. UN *Indicators for Sustainable Development*, UNCHS *Urban Indicators Programme*, WHO *Catalogue of Health Indicators*).
Sources of further information	Give full details of references and other sources of information relevant to the indicator (e.g. Web addresses, databases). List, in particular, references to other indicator sets using similar indicators, examples of the use of the indicator, or materials which describe the context and rationale for its use.
Involved agencies	List agencies which have a leading role in relation to the indicator, including data providers, indicator developers and indicator users. Include international, national and (where relevant) regional/local agencies.

Example indicator

Definition of indicator	Detailed definition of the example indicator.
Underlying definitions and concepts	Definition of all terms and concepts involved in describing and constructing the example indicator.
Specification of data needed	List data needed to construct the indicator.
Data sources, availability and quality	Outline potential sources of data, and comment on their quality and characteristics in terms of the indicator. Where appropriate indicate ways of obtaining data which are not readily available (e.g. through special surveys).
Computation	Specify the way in which the indicator is computed, i.e. how the data are analysed/processed to construct the indicator. Where relevant, express the computation process mathematically, and define the terms used.
Units of measurement	Specify the units of measurement used in presenting the indicator.
Interpretation	Describe the ways in which the indicator may be interpreted in relation to the issue(s) specified. Show what inferences can be made from apparent trends or patterns in the indicator. Discuss, in particular, constraints on the interpretation of the indicator, due for example to limitations of the data or complexities in the relationships implied by the indicator.

each indicator. Full profiles for a selection of the indicators listed in Table 4.1 are given in Annex 1.

The examples are important, because they help to emphasise that the "devil" in terms of indicator development is often in the detail. Indicators are only as good as the data on which they are based, and thus issues of data quality are crucial. The way in which indicators are calculated and computed also affects the information they give, and therefore attention needs to be given to the methods used. In addition, indicators are also often used by people who are non-experts in the scientific field. Consequently, the way they are compiled, presented and interpreted is equally important. The use of indicators to communicate information on environmental health risks raises serious ethical considerations (Higginson and Chu, 1991). The profiles presented in Annex 1 thus help to describe environmental health indicators in ways which enable users to understand them better, and to share good practice in indicator development.

4.5   Conclusions

Environmental health indicators serve many purposes and take many forms. To be effective, however, they need to be carefully designed and subject to rigorous quality control. In many ways, indicator development can be likened to an engineering process, in that indicators must be designed to meet the users needs, taking account of the limits and constraints of the available materials and technology (data, models, knowledge, etc.). In particular, usable environmental health indicators depend heavily upon the existence of known and definable links between environment and health. Difficulties in establishing these relationships (due, for example, to the complexity of confounding effects and the problems of acquiring reliable exposure data) inhibit the practical use of many potential indicators and make it difficult to establish core indicator sets.

Environmental health indicators thus have limits, but if used within these limits they can make an important contribution to improved management and protection of public health. Well-designed and well-constructed indicators provide the capability to define more clearly the environmental health issues that need to be addressed, to prioritise these issues, to identify where action can best be taken, to compare the potential cost-effectiveness of different actions and strategies, to assess the effects of past or current action, to define the remaining research needs, and to bring together and inform the various stakeholders involved. The development and use of purpose-designed indicators to meet specific needs therefore remains a priority.

4.6   References

Briggs, D.J. 1995 Environmental statistics for environmental policy: data genealogy and quality. *Journal of Environmental Management*, **44**, 39–54.

Carstairs, V. and Morris, R. 1991 *Deprivation and Health in Scotland*. Aberdeen University Press, Aberdeen.

FAO 1976 *A Framework for Land Evaluation*. FAO Soils Bulletin No. 32., United Nations Food and Agriculture Organization, Paris.

GRID 1999 *Global Resource Information Database – Quarterly Bulletin. Volume 1, Issue 2.*, United Nations Environment Programme, Geneva.

Higginson, J. and Chu, F. 1991 Ethical considerations and responsibilities in communicating health risk information. J*ournal of Clinical Epidemiology*, **44**(S1), 51S–56S.

Hope, C. and Parker, J. 1991 A pilot environmental index for the United Kingdom: results for the last decade. *Statistical Journal of the United Nations*, **8**, 85–107.

Inhaber, H. 1976 *Environmental Indices*. John Wiley and Sons, London.

Jarman, B. 1984 Underprivileged areas: validation and distribution of scores. *British Medical Journal*, **286**, 1705–9.

Murray, C.J.L. and Lopez, A.D. [Eds] 1996 *The Global Burden of Disease: a comprehensive assessment of mortality and disability from diseases, injuries, and risk factors in 1990 and projected to 2020*. Published on behalf of the World Health Organization and the World Bank by Harvard School of Public Health, Harvard University Press, Cambridge, Massachusetts.

Newman, P.J. 1992 *Classification of Surface Water Quality. Review of Schemes Used in EC Member States*. Heinemann Professional Publishing, Oxford.

Townsend, P. 1987 Deprivation. *Journal of Social Policy*, **16**, 125–46.

UNDP (United Nations Development Programme) *1990 Human Development Report*. Oxford University Press, New York.

WHO 1992 *International Statistical Classification of Diseases and Related Health Problems, Tenth Revision (ICD-10).* Volume 1. World Health Organization, Geneva.

World Bank 1993 *World Development Report 1993: Investing in Health*. Oxford University Press, New York/Oxford.

Chapter 5[*]

ASSESSMENT OF EXPOSURE AND HEALTH EFFECTS

5.1   Introduction

The goal of environmental epidemiology is to understand the health effects of environmental factors that are outside the immediate control of the individual (Rothman, 1993). As such, it encompasses the processes and effects of exposures to physical and chemical pollutants not only in the open environment, but also in occupational environments, together with the study of the spread of infectious agents through environmental media such as the air, water and food. Moreover, psychosocial factors and the public's perception of environmental health risks are increasingly important in environmental epidemiology.

Typically, the exposures that are beyond individual control affect many people simultaneously. Measurement of individual exposures is thus difficult and costly. As a result, environmental epidemiological investigations often have to rely on the use of existing data, and to analyse these at the aggregate rather than individual level. It is also important to appreciate that epidemiological studies require more than data on exposure and health. Equally important are data on other known or possible risk factors which may confound relationships with the health outcomes of interest. Environmental exposures often have small effects that may be masked or distorted by the effects of confounding. Observed health effects of air pollution, for example, may be confounded by risk factors such as smoking or occupational exposures. Socio-economic factors act as confounders for many environmental health effects. The assessment of effect modification (i.e. the change of the strength of the association between exposure and health outcome according to some other factor) is also important for generalising observed exposure–effect relationships to other populations. In environmental epidemiology, problems connected with ecological analyses (i.e. problems with inference based on grouped data) call for further methodological work. For example, by obtaining individual-level data on the exposure and certain covariates in samples of selected groups, it might be possible to determine the

---

[*]   *This chapter was prepared by T. Nurminen, M. Nurminen, C. Corvalán and D. Briggs*

limits of ecological bias in estimating the health effects (Morgenstern and Thomas, 1993; Prentice and Thomas, 1993).

The health effects of pollutants found in the environment can be divided into two broad classes: acute (i.e. short-term) and chronic (i.e. long-term) effects. Each of these can range in severity from death to minor illness or discomfort. Microbiologically contaminated water or food, for example, can have an intense health effect a few hours after a short exposure, but with no detrimental long-term effects on health. Arsenic in water, on the other hand, may have a severe longer-term effect at low but constant exposure levels (e.g. leading to cancer). Lead contamination provides an example of an environmental hazard which can have either acute or chronic effects. Thus, some pollutants may have an almost immediate effect after exposure and other substances may require accumulation in the target organ before causing any detectable adverse health effects. For some pollutants there may be a threshold level, below which no health effect is evident. For others there may be no threshold, and some effect may occur at even the lowest exposure levels. Moreover, some health outcomes may require a period of latency before the effect is observed.

People are not affected equally by the same environmental hazard. Substantial variations in sensitivity to an exposure may thus occur within a population. These differences may derive from a number of factors, including differences in characteristics of the individual. In this context, recent advances in the understanding of the role of genes has been particularly important, although problems may exist in determining whether the marker for sensitivity being examined is a measurement of the genotype itself, some host characteristic, or family history (Hatch and Thomas, 1993).

Age, nutritional status and state of general health are also important determinants of individual vulnerability. Exposure hazards for the normal "healthy" population, therefore, do not necessarily apply to all sectors of the population, and separate assessments may need to be made for particular high-risk groups such as infants and young children, the elderly, pregnant women and their foetuses, the nutritionally deprived, and individuals suffering from some diseases (de Koning, 1987). Such groups can often be identified by assessing the degree of effect modification which occurs for each specific group compared to the "normal" population. It is especially important to identify these high-risk groups because they will usually be the first to experience adverse health outcomes as the level of the pollutant increases. A study in Romania, for example, showed higher values of lead in blood in children living near a lead smelter than in adults in the same area; these results indicated that biomonitoring should be extended at least to children in other parts of the city concerned (Verberk et al., 1992).

Vulnerability of populations to hazards is also evident in the different abilities of individuals or groups to mitigate their exposure to, and the effects of, environmental threats. For example, if microbiologically contaminated water leads to cases of morbidity, the effect of the contamination will depend in part on the ability of an individual or a group to gain access either to alternative water sources or to therapeutic treatment. Ability to cope with the effects of environmental hazards is very often limited by economic circumstances. Thus, while high rates of mortality in an area may in part be due to the existence of an environmental hazard, they are not necessarily a direct indication of pollution levels. Instead, the severity of the effect of the hazard may be more closely related to variations in the ability of individuals to protect themselves from exposure or to treat the effects of the hazard.

5.2   Exposure patterns and processes

The assessment of exposure is clearly fundamental to environmental epidemiology, and methods of exposure assessment have consequently been the focus for much attention in the literature (e.g. AIHA, 1988; ACGIH, 1989; HSE, 1990, 1991; CEN, 1991, 1992; Rappaport and Smith, 1991; Hawkins *et al.*, 1992; ISO, 1992). However, environmental exposures can occur in many different ways. Exposure may take place as a result of inhalation, ingestion or dermal absorption of pollutants which have been carried or stored in the air, water, food, biota (vegetation and animals) and soil. In many cases, exposure may occur simultaneously from many sources and through multiple routes. Pathways of exposure to lead, for example, include air pollution from traffic and industrial emissions, drinking water, food, tobacco smoking, dusts, paints and other industrially produced commodities and soil (IARC, 1982). Valid exposure assessment therefore typically requires detailed knowledge about the geographical distribution of the pollutants of concern, the temporal variations in pollution levels, the processes of exposure, and pathways of exposure, and the time activity patterns of the exposed individuals.

5.2.1   Geographical distributions

The geography of environmental contamination is complex. Different pollutants may be derived from a wide range of different sources, including localised point sources (e.g. industrial chimneys), line sources (e.g. roads) and diffuse sources (such as agricultural activities). Release from any of these sources may also occur either through controlled pathways (e.g. from a stack or discharge pipe) or as fugitive emissions which leak inadvertently into the environment. Once in the environment, they may be transferred by many different processes and pathways. On the way, they undergo a great variety of changes as the result of dilution, deposition, chemical reactions and physical decomposition. Rates of these processes depend upon the pollutant species and the environmental medium concerned. As a result, patterns of pollution differ markedly in their magnitude and extent. Some pollutants may be widely and relatively uniformly distributed, due either to the ubiquitous distribution of their source activities, or the effects of active mixing or long-distance transport. Other pollutants show more localised patterns, reflecting the localised distribution of emission sources and the limited extent of transport. Atmospheric pollutants emitted primarily from tall stacks (e.g. sulphur dioxide from power stations and other major combustion plants), for example, may be widely dispersed. Nitrogen dioxide, which is derived primarily from low-level traffic sources, often shows marked variations even within an individual street. Nitrate and phosphorus pollution of surface waters is extremely extensive. Organic pollution of drinking water, in contrast, commonly occurs at the level of a neighbourhood or household. Food contamination can be specific to a particular product and affect all population groups consuming the product, or it can be specific to a household or neighbourhood where food storage hygiene is locally inadequate. Exposure to electromagnetic fields can vary strikingly over short distances.

5.2.2   Temporal variations

Temporal variations in pollution levels are equally important. Pollution levels typically show a number of different trends at different temporal scales. In many cases long-term trends exist, reflecting underlying changes in the rates of emission (e.g. as a result of technological or economic changes or due to policy intervention). Superimposed upon these there may be annual variations, reflecting year-to-year differences in climate or source activity. Many pollutants also show marked seasonal, weekly and diurnal patterns, due to cycles of activity and short-term climatic and other effects. Major, short-term pollution episodes may also occur as a result of sudden, accidental releases. Measurements of exposure will therefore vary according to both when sampling is carried out and the duration of individual measurements (the averaging time). There are also many different ways of expressing the exposure level, e.g. as the average, peak, percentile (95 per cent and 98 per cent are often used), frequency of exceedance of a specified level, or cumulative duration of exceedance. The time scale of interest and the specific indicator to be used will depend on the health outcome that is to be studied and existing aetiological knowledge about the exposure–effect process. Because of the effects of latency in many health outcomes, the timing of sampling may also need to vary. Concurrent exposures are not always the main concern, but information on past exposures may also be needed — in the case of outcomes such as cancer, information may be needed for many years previously.

In order to model past or future concentrations, or to isolate the effects of specific pollution episodes, it may be necessary to unravel the effects of these different components of temporal variation. This is often extremely complex, for the different cycles are not easily identifiable and are often masked by considerable random variation in pollution levels. Time series analysis is often used for this purpose, but even this method must be used with care because it involves a number of assumptions and decisions on the part of the user which may significantly affect the results.

5.2.3   Measurement issues

Variations in individual absorption or metabolism of pollutants are also complex. Exposure assessment and dose estimation thus pose difficult problems for those investigating the health effects of environmental agents. Concepts of exposure were discussed briefly in Chapter 3. As noted there, the term "exposure" refers both to the concentration of an agent at the boundary between an individual and the environment and to the duration of contact between the two. Dose, in contrast, refers to the amount actually deposited or absorbed in the body over a given time period (Hatch and Thomas, 1993). Internal dose is the ideal measure from the scientific standpoint, but limits and standards set by health and safety legislation usually relate to external exposures. Occupational exposure to lead, for example, is regulated and monitored on the basis of blood lead levels in workers. Blood lead, however, is inadequate either for monitoring organic lead compounds or as an indicator of amounts of lead in target tissues and temporal variations of exposure levels (Kazantzis, 1988). While there is undoubtedly a need to improve externally derived measures of exposure, efforts are also needed to estimate internal dose using methods such as empirical dosimetric modelling, pharmaco-kinetic modelling and biological markers (Hatch and Thomas, 1993).

The long latency time likely to occur between exposure and presumed health effect in many cases further exacerbates the difficulties of exposure assessment (Rothman, 1993). In these circumstances there is a need to link data on present-day health outcomes to data on past exposures. Estimation of past exposures, however, is often exceedingly difficult. Where good historical records are available, it may be possible to make generalised esti-mates of exposures, and examination of past patterns of pollution can provide a basis for modelling (Hatch and Thomas, 1993). Nevertheless, suitable historical data on exposures are often lacking. Changes in pollution levels, place of residence and lifestyle may also mean that it is not realistic to extrapolate back from recent data. For unrecorded and imperceptible exposures (such as electromagnetic fields in particular) retrospective evaluation can be only approximate, at best.

5.2.4   Pollutant combinations

People are often exposed to different pollutants simultaneously. To isolate the effect of one requires that the others have been controlled for in the analysis. Exposure to these may occur at different locations (e.g. in the workplace and/or at home) and at different times. Thus, it may be necessary to establish different sampling regimes or to use different sources of information to obtain exposure estimates.

The full range of factors which may need to be examined in any particular study is therefore potentially large. It may include many different environmental pollutants (including hazardous chemicals, radioactivity, dusts and particulates) from many different anthropogenic (including energy production, industry, pesticide use, transportation, etc.) and natural sources (e.g. geological release of radon), released either continuously or sporadically, and either under controlled conditions (i.e. deliberate discharges) or accidentally. Data on these pollutants may need to be obtained either from monitoring sites within or around the study area, or through the use of modelling techniques. In the latter case, additional data may be needed on levels of source activity (e.g. traffic density, industrial production), emission rates, meteorology and other factors which affect dispersion processes (e.g. topography). Different data sources, sampling regimes and analytical procedures may be needed for the different pollutants and sources involved.

5.3   Sources of exposure data

5.3.1   Exposure sampling strategies

Information on geographical variations in pollution levels may be derived from a number of sources. Often the most useful are the results of monitoring exercises. Almost all countries now run routine monitoring networks for a wide range of pollutants, and networks in many countries are being extended. New sampling and analytical techniques are being developed, including the use of automatic samplers and remote sensing. The development of low-cost sampling devices (e.g. passive samplers) for an increasingly wide range of pollutants is also facilitating the use of purpose-designed surveys. Gradually, improved awareness about the spatial and temporal variations in pollution is contributing to improved sample designs, so that monitoring is being undertaken more effectively (for example, by sampling the micro-environment where exposure principally occurs) including indoor environments (such as bedrooms and living rooms in the assessment of radon and electric and magnetic fields). The use of total exposure monitoring, in which all potentially relevant micro-environments are sampled, also offers opportunities to improve exposure estimates (Hatch and Thomas, 1993). In addition, personal exposure monitoring is being incorporated to some extent into environmental health assessment. The questions that form the basis of any sampling strategy include (Gardiner, 1995):

- What should be measured.
- How the sampling should be done.
- Whose exposure should be measured.
- Where the sample should be collected.
- When measurements should be made.
- How long sampling should go on for.
- How many measurements or readings should be taken.

Nevertheless, in many cases, it is not possible to obtain information directly on pollution levels for the locations or areas of interest. In these cases, models may have to be used to estimate exposures. Several approaches are available. Where suitable data exist, it may be possible to estimate pollution levels by interpolating data from nearby monitoring sites. With the development of GIS, a wide range of interpolation and mapping methods have become available (Briggs and Elliott, 1995). This approach is normally only feasible, however, where the distances involved are relatively small and the spatial variation in pollution levels is limited. Alternatively, it may be possible to estimate concentrations in the areas of interest by using dispersion models. Again, a wide range of models have been developed over recent years, but their applicability is often limited by their relatively stringent data demands. Most air pollution models, for example, require detailed data on emission sources and rates and meteorological conditions (Henriques and Briggs, 1998). Where neither of these methods are possible, it may be appropriate to use more empirical methods. Multiple regression techniques, for example, may be used to construct predictive equations based upon environmental factors thought to determine pollution levels (Briggs *et al.*, 1997).

Obtaining reliable estimates of exposure at the individual or group scale is nevertheless extremely difficult, especially where routinely collected data are being used and variations in concentration are localised. In these cases, the measurement stations may not be representative of the environment in which exposure occurs. Within a geographically defined population, considerable variations in exposure may also occur, reflecting local variations in pollution level and individual behavioural patterns. The application of a single exposure score to the entire group, based for example on results from monitoring stations, is therefore likely to be erroneous and must always be undertaken with care.

One way of improving exposure estimates is thus to take account of people's movements through, and residence times within, the pollution field.

This may be achieved either by using data on time activity patterns collected from purpose designed surveys of the target population (Silvers *et al.*, 1994; Farrow *et al.*, 1997) or by modelling time activity patterns (Ott *et al.*, 1988).

5.3.2   Routinely collected environment data

Routine environmental monitoring provides one of the most important sources of exposure data. Most countries now undertake routine monitoring, and a number of international monitoring networks also operate. The advantages of routinely collected data are that they are likely to be relatively easily accessible (often through government departments) and widely available; to follow approved methods; and to be available for a relatively long period of time. Nevertheless, routine data may not be optimal for exposure assessment and linkage with health data. Problems may include the relevance of the monitoring with regard to the population and the environmental health problems encountered, the frequency of the measurements, the spatial representativeness of the monitoring sites and the geographic and temporal completeness of the data. Examples of such problems include:

- Environmental data may be collected in areas which do not correspond to where the main exposures occur, or to where people live.
- Exposure data relevant to some important environmental health problems may not be collected.
- Data may be collected on a weekly or monthly basis, when more frequent data would be preferable, or data may be recorded on a more frequent basis but only summary data may be made available.
- Data may be collected for certain periods of the year (e.g. when exposures are assumed to be higher), but the excluded data would be relevant for comparison purposes.

Not all the data required can be obtained from government departments. Therefore additional information will often have to be sought, for example, from industry or private research establishments. Typical examples of the data which may be available from these sources include information on the types of pollution and waste treatment and control equipment, details of the manufacturing processes, raw materials used, sales, and data on emission rates. Difficulties with these data sources may include the confidentiality of the data, costs of data acquisition (there is an increasing tendency by many organisations to charge commercial rates for data), and lack of comparability.

5.3.3   Previous field studies

Data may be obtained, in some cases, from the results of previous field studies. These are often conducted as part of research projects or as pilot projects for longer-term monitoring exercises. Large numbers of these studies have been carried out, especially in more developed countries.

Commonly, they cover a restricted geographic area, but within these areas they often involve extremely detailed investigations. For this reason, they can be a rich source of environmental data. Problems may occur, however, in gaining access to results from such studies, because they may not be widely reported, and contacts with data holders may not be easy to arrange. In addition, they may not have been carried out specifically to investigate links between environment and health, and thus the survey design may not be optimal for such applications. The fact that the data are not routinely maintained may also mean that they become out of date quickly. Problems of comparability may also occur where there is a need to combine the data with results from other sources.

5.3.4 Purpose-designed surveys
For the reasons mentioned above, purpose-designed surveys would seem to provide the ideal source of data in many instances. These *ad hoc* surveys have the major advantage that they can be designed specifically to meet the needs of the study, and the sampling framework, choice of exposure indicators and analytical techniques can all be optimised. In practice, however, they have two major drawbacks: they are likely to be costly and time-consuming. In optimising the survey design to meet the immediate needs of the study, comparability with other data sources may also be sacrificed. Furthermore, the short-term nature of most surveys of this type means that their results may become redundant rather rapidly. For these reasons, purpose-designed surveys should normally be undertaken only as a last resort, i.e. when suitable data are not available from other, existing sources. In these circumstances, use of rapid survey techniques and low-cost sampling devices may help to minimise costs and time-delays (WHO, 1982; Economopoulos, 1993).

5.3.5 Finding environmental data
The fact that environmental data are often collected not by official agencies but by private organisations and research groups means that searching for data can be a lengthy and frustrating task. This is especially true where data directories or metadatabases, listing and describing available data sources, do not already exist. Even where directories are available, they may not be sufficiently informative, because they do not necessarily record details of data characteristics, such as the method of georeferencing, spatial resolution and averaging time, all of which may be crucial in determining the suitability of the data for environment–health linkage studies. In the absence of such directories, data availability can often only be established through direct contacts with potential data holders and by careful literature searches.

In recent years, several developments have occurred which have begun to enhance access to data. One is a general improvement in the recording and reporting of data, often as part of Internet services. Associated with this, many agencies are now making their data publicly available via the Internet. A third development has been the work of national and international agencies to collate environmental data and establish central data repositories, archives or databases. One international example is the WHO Healthy Cities Air Management Information System (AMIS), which provides data on monitored concentrations of air pollutants, worldwide.

5.4   Environmental data quality

5.4.1   Problems in environmental data
The complexity of the environment, the high costs of monitoring, and the technical limitations of many environmental monitoring techniques, mean that environmental data are subject to severe problems of quality. Major problems typically include:

- Gaps in data coverage and completeness due to, for example:
  - equipment failure;
  - detection limits (e.g. use of equipment which is unable to detect low concentrations of pollutants);
  - failure to report or analyse data;
  - gaps in the sampling network;
  - cessation of sampling programmes; or
  - disruptions such as war, strikes or storms.
- Lack of data comparability due to, for example:
  - changes in measurement techniques;
  - changes in sampling design;
  - changes in analytical, classification or reporting methods;
  - changes in the parameters measured; or
  - administrative changes (e.g. in the administrative units for which data are collected).
- Bias and error due to, for example:
  - non-representativeness in the sample design;
  - measurement error (e.g. poor detection);
  - analytical or modelling error;
  - reporting or transcribing error; or
  - aggregation error (e.g. rounding).

The effect of all these factors is to introduce considerable uncertainty into many environmental data sets (Briggs, 1995; Elliott and Briggs, 1998). In the case of atmospheric emissions, for example, it has been suggested that current techniques may have potential errors ranging from about 10 per cent for SO_2 to 100 per cent or more for volatile organic compounds, due primarily to uncertainties in the emission factors and source activity data used. Moreover, changes in the emission factors used mean that emissions data are often recalculated. In the UK, for example, there was a 40 per cent change in the estimates of annual NO_2 emissions between 1983 and 1992, due to adjustments in methodology (Briggs, 1995).

As noted previously, national air pollution networks are often too sparse to detect local variations in pollution concentrations. Similarly, many national monitoring networks for stream-water quality collect samples on only a few occasions each year, so they give only poor estimates of the annual pollution level and provide little or no data on short-term variations. Estimates of waste generation and collection are typically based on only the most limited monitoring and face severe problems of how to classify and quantify waste materials. As a result, estimates may have margins of error considerably in excess of 100 per cent (Briggs, 1995). For all these reasons, environmental data must be treated with considerable caution.

5.4.2 Quality control

In the light of all the problems inherent in environmental data sources, quality control is of the utmost importance. Poor quality exposure data can totally undermine attempts to analyse linkages between environment and health. It is therefore vital to have good knowledge of the data collection procedures, so that the reliability of the data can be assessed (and, if necessary improved). This is particularly important where data were originally collected for purposes other than exposure assessment. Unfortunately, there is generally a lack of supporting information on the genealogy of environmental data sets. It is also often difficult or impossible to obtain independent measures of pollution or exposure against which to verify the data being used. As a result, it is often difficult in practice to check the quality of the data. There is an urgent need to establish standards for reporting and documenting data definitions and genealogy. In addition, the techniques available for quality assessment are as yet poorly developed. Equally important, therefore, is the development and application of improved methods for assessing and reporting data quality in environmental epidemiology (Hatch and Thomas, 1993).

It is particularly crucial to check the quality and consistency of information where data are obtained from different sources, because otherwise inherent inconsistencies may be overlooked. Among others, the following techniques can be used:

- Constructing scattergrams to examine the relationship between exposure indicators and to search for obvious outliers.
- Visually comparing data with other, independently published sources.
- Statistical comparison of data from different sources.

- Mapping individual indicators or use of trend surface analysis techniques to look for discontinuities which coincide with the boundaries between different data sources.

5.4.3   Data standards

If valid comparisons between countries or cities are to be made, it is evident that environmental data standards need to be improved. The health-related programmes of urban air quality, water quality and food contamination, carried out under the Global Environment Monitoring System (GEMS), have performed a valuable service in this respect by providing a framework for a standardised system of data collection which countries can follow. They have also provided advice on which exposures to monitor, and encouraged other countries to participate in this worldwide monitoring effort. In addition, the GEMS Human Exposure Assessment Locations (HEAL) programme has provided resources directed to the collation of accurate and reliable data on human exposures (UNEP/WHO, 1993). Within Europe, both the European Environment Agency and Eurostat also have a major role in establishing standards and procedures for data collection, in conjunction with other international agencies such as OECD and the United Nations Economic Commission for Europe (UN-ECE).

Nevertheless, the adoption and implementation of common standards is not always feasible. The historic investment that many countries (especially in the developed world) already have in monitoring systems, for example, may make them reluctant to change to new, international norms. Local or national priorities and circumstances may mean that standards developed elsewhere are not considered relevant. To identify and analyse local problems may require the use of specific methods and indicators. Ensuring comparability with other areas or countries, therefore, is not always appropriate. Nevertheless, much useful information may be lost if cross-comparisons between results from different studies cannot be made (e.g. in order to obtain improved estimates of exposure–effect relationships from a wider range of areas). Even when developing specific, detailed studies, therefore, it is important to bear in mind the potential wider relevance of the results, and to design the study accordingly.

5.5   Health assessment

Adverse health outcomes due to environmental exposures represent a broad spectrum of effects. They range in scale from the population to the individual, and in magnitude of effect from premature death to severe acute illness or major disability, chronic debilitating disease, minor disability, temporary minor illness, discomfort, behavioural changes, temporary emotional effects and minor physiological change (de Koning, 1987).

Traditionally, concern about environmental hazards has tended to focus upon hazards believed to be contributing to excess mortality, in part because of the relative ease of obtaining mortality statistics. Nevertheless, relatively few studies have shown clear associations between environmental pollutants and actual excess in deaths. Even where it would otherwise have been expected, investigation has usually revealed no evidence of gross excess mortality (Lancet, 1992). The main exceptions are serious accidents or events which have resulted in release or accumulation of large amounts of toxic substances in the environment, leading to deaths due to poisoning. Long-term effects on mortality are invariably even more difficult to demonstrate. Typically, only a small subset of the population experiences high levels of exposure, and the doses received by the general population are so low that only vulnerable high-risk groups are severely affected. As a consequence, any excess mortality due to a pollutant is restricted to a small section of the population. Mortality across entire populations thus tends to be a weak and insensitive indicator of environmental health effects in most situations (Landrigan, 1992). Whether mortality is a reliable environmental health indicator, and if so for what groups, must be considered in the context of the particular circumstances.

Because of the general insensitivity of mortality, and because it would also be beneficial to detect the effects of exposure long before death, there are obvious advantages in having other, earlier measures of health outcome. One way of doing this is to use data on morbidity. In some cases this is relatively straightforward, especially where formal disease registers exist, such as for cancers (Draper and Parkin, 1992; Swerdlow, 1992). Otherwise, however, obtaining suitable data poses severe problems, due to the inadequacies of many health surveillance and recording systems and the inconsistencies inherent in the data. Disease occurrence, for example, may be measured in many different ways: as number of hospital admissions, length of hospitalisation, drug sales, medical consultations, days-off-work, etc. Each of these measures different components of morbidity and each is subject to substantial differences in reporting rates. Disease prevalence may be influenced by variations in the duration of the disease and survival rates. Incidence data are generally less easily accessed and can be subject to artificial variations in ascertainment (e.g. as a result of screening programmes). In order to avoid dilution of weak associations through inclusion of irrelevant cases, therefore, it may be desirable to focus attention on subgroups of disease which, on the basis of prior observation, can be considered specifically responsive to the exposure of interest (Hatch and Thomas, 1993).

Various more subtle indicators of health outcome may also be sought, such as reproductive and developmental outcomes or premorbid changes in the state of health. Routinely collected data on these effects are rarely available, and reliable data on baseline rates and normal ranges for subclinical endpoints are often lacking (Hatch and Thomas, 1993). Questionnaires can provide an effective means of obtaining data on perceived health, but severe problems may occur in obtaining unbiased response rates across all sectors of the population. Biochemical or physiological changes in individuals, or complaints to local health authorities regarding nuisance factors in the environment, may also be used as outcome measures. Whether these are considered as valid indicators of adverse effects, however, depends on the accepted concept of the term "state of health" (de Koning, 1987). Despite the understandable desire to use earlier indicators of health effect, therefore, serious problems remain in obtaining the relevant data.

5.6   Health data

Health data are clearly of primary importance in environmental health studies. In the context of the HEADLAMP approach, they perform two main roles. Firstly, they provide indicators of the effects of known exposures to environmental pollution on human health. As such, data on health outcome, when linked to appropriate environmental data, can be used to assess or confirm exposure–effect relationships within the study area, or to quantify the contribution of specific exposures to total mortality or morbidity. Similarly, monitoring of health outcome can show the effects of changes in exposure, due for example to policy interventions or the adoption of new technologies. Additionally, they can provide an indication of the possible existence of previously undetected exposures. Thus, variations in health outcome may be used to infer the existence of underlying variations in exposure which need further investigation.

Like environmental data, health data may come from a variety of sources, including routine monitoring, *ad hoc* surveys and purpose-designed studies. These provide data on a variety of indicators, including health status (e.g. infant mortality, progress in child development, blood pressure), disease (morbidity, hospitalisation, incidence or prevalence of different signs and symptoms) and adverse effects (e.g. premorbid changes in the state of health and complaints to local health authorities regarding nuisance factors in the environment).

Results from routine health monitoring programmes might be expected to provide the most appropriate source, because they tend to be available on a continuous basis for the whole of the area concerned, to be relatively easily accessible (at least at an aggregate level), and to be standardised in terms of procedure. Routine monitoring of health is undertaken for a variety of purposes: to provide management information on the performance of the health service, to monitor trends and detect changes in health status, to provide an early warning about health problems, and to monitor the need for and effects of health policy. It is these requirements, rather than any explicit need to link the health data with information on the environment, which consequently determine the design of the monitoring systems. As a result, routine monitoring does not necessarily provide ideal data for environment–health linkage studies. Moreover, like all health data, routinely collected information may be subject to errors and inconsistencies in diagnosis, reporting and georeferencing.

5.6.1   Mortality data

Data on causes of death are available in most developed countries and are the only health statistics for which comparatively long time series are available. Variations in diagnostic practice and coding will, however, affect the comparability of death certificate information between different regions within a country or between countries. Cause-specific mortality data may also be subject to misclassification. Each year, WHO receives mortality data classified according to cause from 37 developing countries and about the same number of developed countries. This information is readily available from WHO and is published yearly in the World Health Statistics Annual Report. Of the developing countries, only 22 consider that the reporting of deaths is complete (WHO, 1987). Therefore, very few developing countries are in the position to monitor changes in causes of death on the basis of complete and reliable data.

Data on infant mortality are considered to be an indirect indicator of the level of health in the population. There are, however, a number of conceptual and practical problems with this indicator. A particular problem relates to differences in the definition of "infant death" for registration purposes in the first few days of life. The coverage of countries and areas in developing regions of the world in which registration of infant deaths is at least 90 per cent complete is much less than for those reporting total population births and deaths (United Nations, 1985).

Many studies of environment–health relationships rely on time series analysis. These require short-term (e.g. daily) counts of mortality. Daily mortality data are likely to be available in many countries, but perhaps not always in a form that is useful for computer analyses. Extra data entry or data processing may therefore be required.

5.6.2   Morbidity data

Morbidity statistics are generally less readily available than mortality data even for developed countries. Typically, they are less complete and often refer only to specific subsections of the population. One exception to this is data on infectious diseases of significant public health importance. In most countries these must be recorded, and their reporting to a central health authority is often a legal requirement.

The accuracy of morbidity information depends on a number of factors, including the extent to which patients seek and obtain medical help, diagnosis practice and accuracy, the notification procedures, and treatment procedures. Variations in morbidity, therefore, do not necessarily reflect underlying differences in risk. When considering a small area, for example, it may be difficult to conclude whether a high prevalence of a disease is due to a poor immunisation rate or a good reporting of cases. Ongoing monitoring, instead of a cross-sectional assessment, is therefore desirable.

Disease registers are useful for obtaining incidence data on specific conditions. Most countries have registers of diseases, in particular of cancer. Other well organised registers include those of congenital malformations and mental disability. The usefulness of disease registers depends upon their level of completeness and the quality of their records. Good registers may reach 95 per cent completeness or greater, but there may be significant unevenness in the level of completeness between areas (Swerdlow, 1992). Unfortunately, independent data against which to assess the completeness of disease registers are rarely available, although indirect measures (such as mortality to registration ratios) may be used to indicate discrepancies (Muir and Waterhouse, 1987). A register of a terminal disease, such as cancer, may not be considered complete until data from death certificates are used to complement those from referring hospitals and other regular sources. Other problems include duplicate registrations, differences in practice for dealing with multiple cancers, methods of georeferencing, and delays in registration (Swerdlow, 1992). Moreover, not all registers are yet fully computerised. A considerable investment of resources may thus be necessary to capture the data in a form suitable for analysis.

Annual data on cancer incidence are reported to the International Agency for Research on Cancer (IARC) from registries in participating countries. Although more than 20 developing countries report to IARC, data refer to population subsections and in many cases there is a question as to their reliability. Some countries keep specific registers for certain diseases, such as myocardial infarctions or congenital malformations. In other cases, information on specific diseases is collected through purposely-designed health surveys of representative populations or specific high-risk groups.

Information on communicable diseases is also available in many countries, and routine monitoring has played an important role in disease control in developed countries. Data on these diseases may be collected in a variety of ways, including mandatory notification, surveillance, sentinel networks and laboratory networks. The task of assessing the health impact of different communicable diseases on the population is made easier, in many cases, because (in contrast to chemical and physical agents) the health effects tend to be very specific for a particular exposure (e.g. hepatitis caused by hepatitis virus).

Monitoring of occupational diseases and accidents has proved to be effective in their prevention. As a result, most industrialised countries have established monitoring programmes for occupationally exposed populations, while developing countries are in the early stages of implementing such programmes. In recent decades, increasing numbers of countries have also linked mortality data with occupation and place of residence. This has brought to light several associations with potential aetiological factors, although in most cases subsequent epidemiological analyses have been required to confirm the relations. Linkage of mortality data from health registers with exposure data can further enhance the detection of environmental risk factors. The effectiveness of such monitoring is increased with diseases specifically caused by environmental factors, such as pleural mesothelioma, lung cancer and asbestosis caused by the inhalation of asbestos dust. Similarly, the linkage of mortality and incidence data from cancer registers with information on occupation has provided a great deal of information on occupational cancers.

Additional sources of information exist in most countries that can be used for assessing disease and disability levels. These include hospital records, health service files, health insurance and physical payment systems, school records, workday losses, and the sales of pharmaceutical products. Although not ideal, these sources provide the basis for constructing indicators of certain aspects of health. Hospital morbidity data have the advantage of being detailed and fairly accurate, but detailed information is normally not coded, and thus for many applications, data capture can be a time-consuming process. Multiple admissions are also not always easily detected, while differential use of health services is a well recognised problem. Another problem is the difficulty in determining the denominator population for the calculation of rates.

Sources of data, such as hospital admission or discharge records, cancer registers and records of congenital malformations, do not on average meet the same levels of exhaustiveness and standardisation as mortality statistics (except in the Nordic countries). The potential value of these systems is nevertheless considerable, because they offer the opportunity to detect and monitor health effects in advance of mortality. It is therefore extremely important to improve these systems by increasing their accuracy, completeness and accessibility.

Health surveys also provide a valuable source of morbidity data. Surveys are routinely performed in many countries, while special surveys may be undertaken to investigate specific health issues. The usefulness of the survey results depends to a great extent upon the survey design. Many surveys are targeted deliberately at particular sections of the population (e.g. high-risk groups) and thus do not provide data on the health of the general population.

Surveys may also be designed to give only national data; sample sizes may then be too small to provide reliable estimates at the regional level.

Specific local or regional surveys may also be carried out to supplement existing data. A detailed survey of the region of interest, for example, may be the best means of obtaining detailed morbidity data for a city or region. A census of a region's hospital or any other health agency will provide information on the major reasons for service utilisation.

The two main survey designs are the cross-sectional and longitudinal survey. A cross-sectional (prevalence) survey is often the most practicable, because it provides a picture of the population at one point in time, making it a rapid and inexpensive method. Longitudinal surveys collect information over time, providing a useful moving picture of the population (measuring change of health status), but at considerable expense and requiring a long duration.

The quality of the results again depends upon the sample design. Probability sampling is often the most reliable way of ensuring that the survey can provide valid sample-to-population inferences. If the region to be surveyed is very large, areas within the region can be selected randomly using, for example, a multistage or stratified sampling technique. Determination of sample size is of great importance because it limits the precision of the survey estimates and constrains the analyses that can be legitimately carried out. The great advantage of a survey is that it can be designed to meet the specific needs of the study. Thus, it can be as detailed as necessary, and information on all the indicators of interest (e.g. morbidity, risk factors and population characteristics) can be obtained simultaneously and within a consistent sampling framework. Although most surveys are designed to obtain data on morbidity, mortality data can be estimated by asking interviewees about deaths in the family. This is particularly useful, for example, in estimating infant mortality.

General guidance on survey methodology can be found in textbooks on sampling techniques (e.g. Cochran, 1960). A number of specialised publications are also available on survey sampling methods for the assessment of human health (e.g. Lutz *et al.,* 1992).

5.7   Population and covariate data

Interpretation of patterns in health outcome cannot be carried out reliably without reference to the underlying population or to variations in those factors which may act as potential confounders to the relationship between environment and health. For these reasons, most studies of environmental health rely on the availability of data on population and covariates, such as social conditions and lifestyle. Moreover, processes such as in- and out-migration create major difficulties in interpreting exposure–health relations on either a temporal or spatial basis (Hatch and Thomas, 1993).

5.7.1   Population data

Data on population numbers are essential for most environmental health studies. Expressed merely in absolute terms, data on health have little meaning, because variations are likely to depend more on differences in the size of the population than on any underlying differences in health. For most purposes, therefore, it is more appropriate to express health outcome as rates — and this requires data on the population as a denominator. In some cases, simple population totals (by gender) may suffice for this purpose. These data are normally readily available from censuses, at least at the national level. Of the 218 countries or areas from which the United Nations Statistical Office requests demographic data, only 15 have not reported an official estimate since 1979. Nevertheless, most countries carry out complete censuses on about a ten-year cycle and thus, at any one time, population statistics may be considerably out of date. Therefore, estimates are generally made, based on population projections. Although these may be reasonably reliable at the national scale, considerable errors may develop over time at the small-area scale. Projections also tend to become less reliable with increasing time since the base census was conducted. Moreover, errors in enumeration are common in censuses, while significant differences may occur in the definitions of the resident population between different countries (e.g. in how transients are classified). Because these errors and discrepancies often affect specific sections of the population disproportionately, population data for certain social or age-groups may be particularly vulnerable to uncertainty.

For many applications, data are needed not merely on total population, but on population subgroups (e.g. by age and gender). These are necessary, for example, where health effects are being studied within a specific age group (e.g. children), where disease rates may vary substantially between different ages and genders, or where time-trends are being analysed. For this purpose, vital statistics are ideally required. These provide a demographic profile of the population under study, which is essentially a count of persons cross-tabulated by age and sex and other personal characteristics. This information allows a computation of standardised rates as a basis for comparison both of the same population at other points in time, and with other populations.

To some extent, this information can be obtained from national censuses. While population by age is widely available for most developed countries, the number of developing countries for which reliable periodic estimates are available is much smaller than those with total population counts. Typically, data on population age structure are only available for census years, and the age classes used in different countries may differ, so that international comparisons may be difficult. Nevertheless, most countries also maintain some form of vital statistics which include registration of births and deaths.

Globally, reasonably complete registration of births and deaths occur in about 85–90 countries or areas (United Nations, 1985). These include all the developed countries and about 40 developing countries. In about 60 developing countries, the registration of vital events is considered incomplete (WHO, 1987).

5.7.2   Confounder data

As already noted, control of confounding is an important element of most ecological studies. Rarely are the relationships between environment and health simple and unitary; instead, they are usually affected by a variety of confounding variables, many of which are only partially known. Rarely, therefore, will interpretations of the linkages between environment and health be wholly valid unless allowance for potential confounding is made.

The confounder data required will clearly depend on the specific relationship being studied. A wide range of potential confounders may exist, including social factors (e.g. ethnic origin, occupation, housing condition, income, education), lifestyle (e.g. diet, smoking, drug use) and physical environment (e.g. exposure to other pollutants, climate). Obtaining data on these confounders is often one of the most difficult aspects of ecological studies. Some data may be available from routine sources, such as censuses and lifestyle surveys, but the scope of these is often severely limited. Data may also be obtainable from attitudinal surveys and market research studies. With the growing opportunity to use such information for the targeting of advertising and direct sales operations, a growing number of databases are being compiled. They can provide useful information on a wide variety of social and lifestyle factors, including diet, income, housing status, smoking, household size and leisure patterns. They may, however, be relatively costly to acquire and data quality may be uncertain. In addition, the possibility exists to acquire data on confounders through purpose-designed surveys. As with acquisition of environmental or health data, these have the advantage of providing better control over the data collection process, and thus ensuring that the data specifically meet the needs of the study. Typically, however, they are expensive to conduct and may cause considerable delay.

Because of the limitations of data availability, it is often impracticable to obtain information on all the confounders of interest. In many situations, therefore, proxies need to be used, based on other, readily available, demographic or social statistics. Most covariates used in ecological regressions are, in practice, either proxies or rather indirect or crude measures of the true confounder. The use of proxies, however, is clearly only valid where they do in fact provide a reliable surrogate for the confounder of concern.

Unfortunately this is not always the case or, at least, the validity of the proxy is a matter of conjecture. In these circumstances, particular care is needed in interpreting the results.

Investigation of the occurrence of lung cancer in cities provides an example. The causes of the higher incidence of lung cancer in many cities are insufficiently known, but are suspected to be related to smoking and socio-economic status (among other factors). A study in Helsinki showed an apparent increase in cancer incidence with increasing mean SO_2 concentration (Pönkä et al., 1993). To interpret this correctly, however, clearly required the possible effects of confounding by smoking and other social factors to be taken into account. Information on smoking habits was not readily available and thus, instead, average education level was included as a covariate in the ecological regression on the assumption that smoking levels were higher amongst the less-well educated. The analysis did, indeed, show a strong inverse association between education and cancer rate. Use of education level in the analysis thus helped to allow for some form of social confounding effect. Nevertheless, to interpret the results as evidence that smoking is related to lung cancer in the study area relies on the assumption that education level is a valid proxy for smoking rate.

The problem of controlling for confounders is further compounded by the potentially large number of confounders that may be of relevance, and the complex relationships that may exist between them. In other words, confounders do not necessarily act individually or in isolation, but may operate in unison. There is consequently a need to measure the multivariate (joint) distribution of the confounders; univariate distributions of the covariates or use of a simple confounder score may not suffice to achieve full control of confounding. Bobak and Leon (1992), for example, carried out an ecological study in the Czech Republic to test the hypothesis that atmospheric levels of pollution affect infant mortality risk. The socio-economic data available included mean income, mean savings, mean number of persons per car, proportions of total births outside marriage, and legally-induced abortions per 100 live births. While these allowed for control of a number of potential confounders, they were clearly not comprehensive, and allowance could not be made for potentially important confounders such as smoking, indoor pollution from heating or cooking, and family size. The potential also existed for interactions between the various confounders. As the investigators themselves acknowledged, therefore, an unknown amount of residual confounding may have been left unresolved. The problem of missing or inadequate information on confounding factors is especially serious in studies using aggregate data.

5.8 Georeferencing

A particularly important need in relation to almost all environmental health data (including data on exposures, health outcomes, population and confounders) is the method of georeferencing. This refers to the way in which individual cases or patients are related to a geographic location. Some form of georeferencing is clearly needed to allow data on health outcome and other factors to be mapped. For this purpose, it may suffice simply to aggregate data to the administrative district or area in which they live. Often, however, it is more appropriate to map health outcomes on an individual level; for example rare diseases such as leukaemias, or communicable diseases where the degree of local clustering may be of interest. In these cases, individuals need to be referenced to a single point location (e.g. their place of residence). Accurate georeferencing may also be necessary in order to enable estimates of exposure to be made. Where exposures do not vary greatly over small distances (e.g. exposures to contaminants in piped water supplies), it may be sufficient to relate individuals to the area or district in which they live. Where exposures vary more locally, however, a higher resolution of georeferencing will be required. Exposures to road traffic pollution, for example, may be estimated by assessing the distance of the place of residence from the nearest main road (e.g. Nitta *et al.,* 1993). More sophisticated estimates of exposure may involve locating the place of residence on a pollution map or by modelling the pollution level at the place of residence (Pershagen *et al.,* 1995). In each of these cases, point locations are required for the individuals in the study, although data may be subsequently re-aggregated to an area basis for further analysis and mapping.

Most health, socio-economic and demographic data have some form of georeferencing. Commonly this allocates individuals to an administrative region or area, based on their place of residence, although some health data may be based on the location of the treatment centre to which the individual is referred (e.g. the hospital). More detailed georeferencing is also possible in many countries using the postal or zip code (where this is available). For an accurate point location, however, information is needed on the address of the place of residence, and a system needs to be available for translating this to geographic co-ordinates. In countries which have cadastres, covering places of residence, this is straightforward. In some countries, GIS-based systems have been set up which provide an automatic conversion between the address and a point location (e.g. the AddressPoint system in the UK). Where such systems do not exist, georeferencing may need to be done manually, and this is a time-consuming process. Problems may also arise in relation to patient confidentiality, which may restrict access to address-based data. Caution is also needed in using any locational data based on the place of residence, because this may not provide an accurate indication of where the exposure occurred. Exposures may take place over a wide area, because people move around the environment. In the case of occupational exposures it is the place of work, not the place of residence, which is most critical. Nevertheless, people may have moved house or job since the exposures took place. Georeferencing thus needs to be carried out carefully, and the method of georeferencing used should be appropriate to the questions being addressed.

5.9 Conclusions

Investigation of the relationships between environment and health, and monitoring of the environmental health situation as part of the HEADLAMP process, both require the ability to analyse and link environmental and health data. Information on exposures may be obtained from a wide range of sources, including ambient or personal monitoring and use of modelling techniques. Routinely collected health data are also often available, especially for mortality and for communicable diseases and cancers. Use of routine health data has many advantages, not least of cost and improved comparability. In many cases, however, health data need to be collected through special surveys and studies. Special care is needed to ensure that such surveys are rigorously designed and provide representative data on the population of interest. Interpretation of patterns in health outcome also requires reference to the underlying population or to variations in other factors which may act as potential confounders to the relationship between environment and health. Data are therefore also required on population and covariates, such as social conditions and lifestyle. In all cases, accurate georeferencing of these data is essential, in order to allow for accurate mapping of the information, and to help draw valid inferences about the relationships between environmental conditions and health.

5.10 References

ACGIH 1989 *Air Sampling Instruments for Evaluation of Atmospheric Contaminants.* 7th Edition. American Conference of Governmental Industrial Hygienists, Cincinnati.

AIHA 1988 *Quality Assurance Manual for Industrial Hygiene Chemistry.* Prepared by the Analytical Chemistry Committee of the American Industrial Hygiene Association.

Bobak, M. and Leon, D.A. 1992 Air pollution and infant mortality in the Czech Republic, 1986–88. *Lancet,* **340,** 1010–4.

Briggs, D.J. 1995 Environmental statistics for environmental policy: data genealogy and quality. *Journal of Environmental Management,* **44,** 35–52.

Briggs, D.J. and Elliott, P. 1995 GIS methods for the analysis of relationships between environment and health. *World Health Statistics Quarterly,* **48,** 85–94.

Briggs, D.J., Collins, S., Elliott, P., Fischer, P., Kingham, S., Lebret, E., Pryl, K., van Reeuwijk, H., Smallbone, K. and van der Veen, A. 1997 Mapping urban air pollution using GIS: a regression-based approach. *International Journal Geographical Information Science*, **11**, 699–718.

CEN 1991 General Requirements for the Performance of Procedures for Workplace Measurements — prEN 482. Prepared by CEN Technical Committee 137, European Committee for Standardisation, Brussels.

CEN 1992 Workplace Atmospheres — Guidance for the Assessment of Exposure to Chemical Agents for Comparison with Limit Values and Measurement Strategy — prEN 689. Prepared by CEN Technical Committee 137, European Committee for Standardisation, Brussels.

Cochran, W. 1960 *Sampling Techniques*. Wiley, New York.

de Koning, H.W. 1987 *Setting Environmental Standards — Guidelines for Decision-making*. World Health Organization, Geneva.

Draper, G.J. and Parkin, D.M. 1992 Cancer incidence data for children. In: P. Elliott, J. Cuzick, D. English and R. Stern [Eds] *Geographical and Environmental Epidemiology*. Oxford University Press, Oxford, 63–71.

Economopolous, A.P. 1993 *Assessment of Sources of Air, Water and Land Pollution. A Guide to Rapid Source Inventory Techniques and their Use in Formulating Environmental Control Strategies.* (2 Volumes), World Health Organization, Geneva.

Elliott, P. and Briggs, D.J. 1998 Recent developments in the geographical analysis of small area health and environmental data. In: G. Scally [Ed.] *Progress in Public Health.* FT Healthcare, London, 101–125.

Farrow, A., Taylor, H. and Golding, J. 1997 Time spent in the home by different family members. *Environmental Technology*, **18**, 605–13.

Gardiner, K. 1995 Needs of occupational exposure sampling strategies for compliance and epidemiology. *Occupational and Environmental Medicine,* **52**, 705–8.

Hatch, M. and Thomas, D. 1993 Measurement issues in environmental epidemiology. *Environmental Health Perspectives,* **101** (Suppl. 4), 49–57.

Hawkins, N.C., Jayjock, M.A. and Lynch, J. 1992 A rationale and framework for establishing the quality of human exposure assessments. *American Industrial Hygiene Association Journal,* **53**, 34.

Henriques, W. and Briggs, D.J. 1998 Environmental modelling in the NEHAP process. In: D.J. Briggs, R.M. Stern and T. Tinker [Eds] *Environmental Health for All. Risk Assessment and Risk Communication in National Environmental Health Action Plans*. Kluwer, Dordrecht, 113–32.

HSE 1990 *General methods for sampling airborne gases and vapours, MDHS 70, October 1990*. UK Health and Safety Executive, London.

HSE 1991 *Analytical quality in workplace air monitoring, MDHS 71, March 1991*. UK Health and Safety Executive, London.

IARC 1982 *IARC Monographs on the Evaluation of the Carcinogenic Risk of Chemicals to Humans. Volume 23. Some Metals and Metallic Compounds*. International Agency for Research on Cancer, Lyon.

ISO 1992 *Harmonized proficiency testing protocol, ISO/REMCO No. 231 Revised, January 1992*. International Organization for Standardization, Geneva.

Kazantzis, G. 1988 The use of blood in the biological monitoring of toxic metals. In: T. Clarkson, L. Friberg, G. Nordberg and P. Sager [Eds] *Biological Monitoring of Toxic Metals*. Plenum Press, New York and London, 547–66.

Lancet 1992 Environmental pollution: it kills trees, but does it kill people? (Editorial). *Lancet*, **340**, 821–22.

Landrigan, P.J. 1992 Environmental pollution and health. *Lancet*, **340**, 1220.

Lutz, W., Chalmers, J., Hepburn, W. and Lockerbie, L. 1992 *Health and Community Surveys*. Volumes 1–2. Published for the International Epidemiological Association and World Health Organization by MacMillan Press Ltd.

Morgenstern, H. and Thomas, D. 1993 Principles of study design in environmental epidemiology. *Environmental Health Perspectives,* **101** (Suppl. 4), 23–38.

Muir, C. and Waterhouse, J. 1987 Comparability and quality of data: reliability of registration. In: C. Muir, J. Waterhouse, T. Mack, J. Powell and S. Whelan [Eds] *Cancer in Five Continents*. Volume V. IARC Scientific Publication No. 88., International Association for Research on Cancer, Lyon, 145–169.

Nitta, H., Nakai, S., Maeda, K., Aoki, S. and Ono, M. 1993 Respiratory health associated with exposure to automobile exhaust: 1. Results of cross-sectional studies in 1979, 1982 and 1993. *Archives of Environmental Health*, **48**, 53–8.

Ott, W., Thomas, J., Mage, D. and Wallace, L. 1988 Validation of the simulation of human activity and pollutant exposure (SHAPE) model using paired days from the Denver, CO, carbon monoxide field study. *Atmospheric Environment*, **22,** 2101–13.

Pershagen, G., Rylander, E., Norberg, S., Eriksson, M. and Nordvall, S.L. 1995 Air pollution involving nitrogen dioxide exposure and wheezing bronchitis in children. *International Journal of Epidemiology*, **24**, 1147–53.

Prentice, R.L. and Thomas, D. 1993 Environmental epidemiology: data analysis. *Environmental Health Perspectives*, **101**(Suppl. 4), 39–48.

Pönkä, A., Pukkala, E. and Hakulinen, T. 1993 Lung cancer and ambient air pollution in Helsinki. *Environment International,* **19**, 221–31.

Rappaport, S.M. and Smith, T.J. [Eds] 1991 *Exposure Assessment for Epidemiology and Hazard Control*. American Conference of Governmental Industrial Hygienists, Lewis Publishers Inc, New York.

Rothman, K.J. 1993 Methodologic frontiers in environmental epidemiology. *Environmental Health Perspectives,* **101**(Suppl. 4), 19–21.

Silvers, A., Florence, B.T., Rourke, D.L. and Lorrimer, R.J. 1994 How children spend their time: a sample survey for use in exposure and risk assessments. *Risk Analysis*, **15**, 931–44.

Swerdlow, A.J. 1992 Cancer incidence data for adults. In: P. Elliott, J. Cuzick, D. English and R. Stern [Eds] *Geographical and Environmental Epidemiology.* Oxford University Press, Oxford, 51–62.

United Nations 1985 *Handbook of Vital Statistics Systems and Methods. Volume 2: Review of National Practices, Series F, No. 35.*, United Nations, New York.

UNEP/WHO 1993 *Guidance on Survey Design for Human Exposure Assessment Locations (HEAL) Studies.* United Nations Environment Programme/World Health Organization, Nairobi.

Verberk, M.M., de Wolf, F.A and Verplanke, A.J.W. 1992 Environmental pollution and health. *Lancet*, **340**, 1221.

WHO 1982 Rapid assessment of sources of air, water, and land pollution. WHO offset publication No. 62, World Health Organization, Geneva.

WHO 1987 *Health Monitoring in the Prevention of Diseases Caused by Environmental Factors.* World Health Organization and Commission of the European Communities, Geneva.

Chapter 6[*]

LINKING ENVIRONMENT AND HEALTH DATA: STATISTICAL AND EPIDEMIOLOGICAL ISSUES

6.1   Introduction

Exploration of associations between environment and health is an integral part of environmental epidemiology, either in the search for previously unknown dose–response relations, or to test hypotheses about such relations. The HEADLAMP methodology is an extension of this approach (WHO, 1995). Lying at the interface between epidemiology and public policy, it involves applying known dose–response relations, established in previous investigations and documented in the literature, to new empirical data as a basis for improved decision-making and policy support.

In general, the data used for environment and health linkage as part of HEADLAMP studies are derived from routine monitoring sources, although where necessary additional data may be collected from purpose-designed rapid surveys. In either case, the data often comprise series of data accrued over a long period of time, and gathered in an aggregated form (e.g. at the small-area or regional level). The need to conduct aggregate data studies arises from the difficulty of acquiring individual-level data, especially on environmental exposures and other covariates (Rothman, 1993). As such, the linkage of a health effect variable (e.g. excess mortality) to exposure and other characteristics of populations does not involve the direct use of individual records. Instead, the HEADLAMP methodology relies on analysing grouped data (Nurminen and Nurminen, 1999).

In the HEADLAMP approach, the aim of the environment and health linkage is not to discover new associations, or to confirm suspected ones. Rather it involves using established scientific knowledge to assess the risks that exist, to identify the need for action, to compare the choices available, and to monitor and evaluate the effects of such actions. As part of this process, the associations previously recognised in environment and health data are extrapolated to new data.

---

*   *This chapter was prepared by M. Nurminen*

The basic method for this purpose is ecological analysis. In addition to the ecological method, however, there is a wide range of more specialised approaches, techniques and procedures which may involve, or be relevant to, environment and health linkage. Examples include the analysis of disease clusters (Rothman, 1990), studies of point source exposures (Elliott *et al.*, 1992), time series analysis (Katsouyanni *et al.*, 1997) and quantitative risk assessment (Nurminen *et al.*, 1999). Each of these may have value in particular circumstances, but each also involves problems and pitfalls about which the investigator needs to be aware. The use of one of the most important tools for exposure and disease mapping, geographical information systems (GIS) (Briggs and Elliott, 1995), is discussed in the next chapter. Together, these methods and tools give an investigator with ingenuity countless opportunities to analyse and exploit existing data at greatly reduced cost. In the process, considerable value is likely to be added to the data, to knowledge about environment–health relations in the area under study, and to the quality of decision-making.

Whatever method is used, if it is to be suitable for linking aggregated environment and health data, two important criteria must be met. First, the method must be simple, inexpensive to implement and applicable to the available data, thus allowing rapid assessment. Second, it must produce statistically valid and scientifically credible results if these are to be used as a basis for action. This means that the method should be unbiased and sensitive to the variations in the data at hand. Ideally, it should yield results that agree with those that would be obtained from more comprehensive *ad hoc* studies (conducted at the individual level) and should provide some estimate of their accuracy and precision.

Section 6.2 below outlines the ecological method in general terms. Section 6.3 reviews time series analysis, which represents a special type of aggregate data method. Section 6.4 discusses the elements of quantitative risk assessment and section 6.5 concludes that the linkage of environment and health data using ecological analysis is useful if used with care.

6.2   Ecological analysis

6.2.1   Background

The basic method for analysing aggregate-level data as part of HEADLAMP studies is ecological analysis. This method involves the investigation of group-level relations between environment and health, by analysing spatial or temporal variations in exposure and health outcome. First used in sociology (Robinson, 1950), it has often been criticised for producing fallacious results. Particular concern has focused on the potential bias which may be introduced by aggregation of data; a problem which Selvin (1958) termed the "ecological fallacy". Despite such theoretical shortcomings, however, ecological analysis has been widely used in environmental epidemiology, not least because it is relatively simple to perform, especially with the large, aggregated databases which are now available. For reasons of logistics and cost it may also be the only approach feasible where large population studies are required. Nevertheless, there has been a growing recognition that ecological or group level associations are not necessarily consistent with those measured at the individual level (Greenland, 1992). Thus, much of the subsequent discussion of ecological methods has focused on how to identify, deal with or avoid the various biases involved, and how to quantify their effects compared with individual level analyses. For the future, more extensive use of the method may be anticipated, stimulated in part by the development of new statistical techniques and GIS.

The ecological approach is a research technique used in observational studies to detect and recognise patterns of disease occurrence across space and time. It is also used to relate the rates of disease frequency to environmental, behavioural and constitutional factors. The ecological design in epidemiology is also useful for the evaluation of intervention on risk factors for various diseases, for example the effect of low-cholesterol diet on the future rate of ischaemic heart disease. Some environmental health problems are more readily approached by ecological studies than by general epidemiological studies. For example, the prevalence of asthma symptoms in relation to climate is applicable to ecological measurement (e.g. Hales *et al.*, 1998), but the occurrence of respiratory symptoms associated with occupational exposure to airborne cobalt is less so. The reason for this is that the level of cobalt exposure of individuals in a worker population is also considerably affected by personal hygiene, because cobalt can absorb through skin, whereas ubiquitous climatic effects afflict a population in the aggregate, rather than as individuals. The ecological method thus derives epidemiological knowledge from the study of disease *of* human populations rather than the study of disease *in* human populations. The grouping variate in ecological analyses is often a geographical region, although other factors such as time period, ethnicity, socio-economic class, etc., could also be used. The situations in which ecological studies are the appropriate design have been summarised in a series of methodological papers summarised in Poole (1994).

Given the availability of suitable exposure and health information, ecological analyses can be conducted in a number of different ways. These can be broadly classified as explorative (disease mapping) studies, multigroup (disease–exposure correlation or regression) studies, or time–trend studies. Disease mapping can detect geographical disease clusters without any direct incorporation of exposure information. The available exposure data allow an epidemiologist to study its association with disease outcome in a single population. Alternatively, one can compare the correlations or, preferably, the coefficients of regression models in two or more populations. In the multigroup design, data on exposure to a risk agent and the health outcome are collected on a group basis for several regions. In either design, the data accrue in a relatively short span of time, but there are typically no multiple measurements over an extended time period. In time–trend ecological studies, a single population may be followed up for changes in exposure over time and the respective changes in the rates of disease over the same period of time.

Ecological analyses of dose–response relations can be potentially biased by several problems (e.g. model misspecification, confounding, non-additivity of exposure and covariate effects, and noncomparable standardisation). Ecological correlations and rate estimates can be more sensitive to these sources of bias than individual level estimates, because ecological estimates are based on extrapolations to unobserved individual level data. In this chapter, emphasis is placed on the biases in group level estimates rather than on their counterparts in individual level designs. Thus the concern here is not so much with the use of the ecological approach in its own right, as with its use as a proxy for individual based studies.

6.2.2   Advantages and disadvantages of ecological studies

Ecological studies continue to be popular because they are often relatively easy to conduct using existing databases in a relatively short period of time. Thus, a judiciously implemented ecological approach can serve as a cost-effective alternative for screening or monitoring of many disease entities and environmental conditions across geographical areas. In practice, however, the true costs of this type of study are often hidden. Establishing and maintaining monitoring systems is expensive, and the apparent cost-effectiveness of this approach only comes from the ability to use relatively low cost, or subsidised, data from a pre-existing monitoring system.

Sometimes an advantage of the ecological approach is that it permits the study of very large populations (e.g. populations of entire countries). Nevertheless, the usefulness of an ecological analysis depends on the purpose of the study, whether it is for basic science, public health, public policy, etc. For scientific studies of disease mechanisms, large populations are not necessarily needed. Moreover population probability samples may offer a better opportunity to study large populations without the limitations of ecological studies. When it is feasible to study large populations ecologically, relatively small increases in risk can be detected. The power of ecological studies, however, is not related to the size of the population studied, but to the number of data points, the accuracy of the data and the power of the statistical methods used. Even so, it is necessary to be cautious about concluding that very small risks are meaningful in practice, simply because they are statistically significantly elevated (see Nurminen, 1997b).

Plummer and Clayton (1996) have studied these important design issues, namely: (a) how large should sample surveys of population exposure be and (b) how should they be targeted on different sections of the study population? They summarised their results as follows "*The number of study populations has little relevance beyond a certain point, the power and precision being limited by the total number of disease events and by the size of the sample surveys used to estimate the distributions of determinants within populations.*" The determination of the optimal size of an ecological study requires also consideration of measurement error, the nature and effects of which are discussed below.

Ecological studies sometimes cover populations more markedly divergent in their exposures than those that can be readily obtained in studies of individuals. Limited within-population variability in exposure may also call for the study of multiple populations in a hybrid epidemiological investigation. For example, when cancer is studied there is the possibility of designed ecological studies in which population exposure is assessed by sample survey methods and compared with reliable cancer statistics. For a discussion of the statistical analysis of such multilevel studies, see Navidi *et al.* (1994) and Sheppard *et al.* (1996).

As noted previously, ecological studies are subject to unique biases not present in individual level studies. Therefore, the demand for methodological rigour is great. The various sources of bias in ecological data derive primarily from linkage failures, i.e. an ecological study does not link individual disease events to individual exposure or covariate data (e.g. see Nurminen, 1995a).

The ecological design provides no information at all on the joint distribution of the exposure and disease variates at the individual level. Thus, there is no way of knowing from the ecological data whether individuals experiencing the health outcome have actually been exposed to the environmental risk factor, or to what level. Inferences on individual level dose–response relations from ecological data are justified only under exceptional, rarely met conditions. Therefore, deriving individual level relations from ecological data should be viewed as a particularly tentative and exploratory process that may yield very tenuous and misleading results.

Problems may also exist with the available data. Routinely registered health event data (e.g. hospital discharges) may not suit the purposes of the ecological research in question, because of an unusable classification system of diseases. It may also be difficult to define the population denominators (e.g. the catchment populations of hospitals) corresponding to the health event numerators. For a less severe health event, such as a mild asthmatic symptom, there may not be any records available at all.

Differences in the geographical basis of the available data may cause additional difficulties. Health data are usually available for administrative units, such as municipal health care districts, municipalities or provinces, whereas data on environmental pollutants and other exposures are often available only for individual monitoring sites or for "natural" areas. Extra effort may be needed to create health and environmental data sets with comparable population subgroups. This may be done either by reallocating individuals to "pollution zones" based on their place of residence or, more commonly, by estimating pollution scores for each administrative area using mathematical models or spatial interpolation techniques. Some of the GIS techniques outlined in Chapter 7 are useful for this purpose.

6.2.3   Biases and problems in grouped level versus individual level studies

In this section, the various sources of bias in ecological data due to invalid study design or incorrect data analysis are considered. These include:

- Aggregation bias.
- Sampling error.
- Measurement error.
- Model mis-specification.
- Confounding.
- Nonadditivity of effects (effect modification).
- Noncomparable standardisation.
- Temporal and spatial problems.
- Incorrect statistical or scientific inference.

Aggregation bias or cross-level bias refers to the incorrect estimates of exposure that result from the analysis of data aggregated across study groups (Robinson, 1950). Because the groups are typically exposed heterogeneously, aggregation bias is a more complex issue than a simple confounding by group (specification bias). A recent attempt to solve this problem has been presented by King (1997). Because the geographical units on which the ecological sampling is based are divisible, ecological analyses are often done at several levels that may not give identical results. Unfortunately, aggregation bias cannot be identified by the examination of results using different aggregations. Ecological results that are similar at all levels of aggregation (e.g. county, economic area, state, region) can still be plagued by aggregation bias.

In addition to the sources of bias ingrained in individual level studies, ecological estimates of effect can be biased from effect modification or confounding by the group variate. Covariates responsible for aggregation bias need not even be effect modifiers or confounders at the individual level (Greenland and Morgenstern, 1989).

The design of the sampling of the population in an ecological study has to account for sampling error (Cochran, 1977). This is a problem that has not been sufficiently touched upon in the literature on ecological studies. When ecological estimates of exposure are based on sample surveys, they are subject to sampling error. If the studies are not based on routine data sources in which sampling errors are negligible (e.g. census data), then exposure variates have standard errors which will bias the regression coefficients. If estimates of the standard errors are available from surveys, these may be incorporated to correct for the bias. However, the sampled areas included in the analysis may differ in size and population density. To allow for the different amount of information contained in each group, a weighted regression should be used with weights proportional to the inverse of the variance of the observation unit.

The susceptibility of ecological estimates to measurement error can be a far more important source of uncertainty than the sampling error. Apart from basic demographic variates (such as sex, age and vital status), most variates used in ecological analyses are measured with error. The design of the information on the samples in an ecological study is more complex than that of classical epidemiological (e.g. cohort) studies. This is because the samples used to estimate the distributions of the disease, exposure and covariate distributions for an ecological study, are often independent. Therefore, the measurement errors that arise from this structure of an ecological study have to be considered separately for the exposure, disease outcome and covariates.

Measurement error has different effects on ecological and individual level studies. Independent non-differential misclassification of an exposure indicator will usually result in a biased estimate of the exposure effect that is directed away from the "no effect" or null hypothesis in ecological studies. Intuitively, this happens because the exposure misclassification reduces the variation in the exposure prevalence across groups; although the group disease rates are unchanged, this effectively magnifies the exposure–effect relation (Sheppard *et al.*, 1996). In individual level studies, random measurement error biases estimates of effect parameters towards their null value and overstates the precision of such estimates (Fuller, 1987).

It is necessary to exercise prudence before drawing firm conclusions about the role of any exposure component because measurement of the exposure level obtained by ecological means may be affected by large random errors. Although it is recognised that such errors tend to bias the observed risk, insufficient attention has been given to the fact that the wrong variate can be identified as the main risk factor (when several variates are correlated with each other and measured with different random errors) simply because one variate may be measured with less error than the others. An analysis of the association of the airborne sulphuric pollutants SO_2 and SO_4 with daily hospital admissions in Ontario furnishes an example of this transfer of causality effect (Zidek *et al.*, 1996). It is suggested that measurement errors be reduced whenever possible by repeating exposure measurements on the same units with different instruments having independent error factors.

Ecological studies in epidemiology typically deal with cause-specific mortality (and morbidity) rates rather than with total mortality. Therefore, misclassification of disease outcome can be a source of severe bias. This bias can be far greater than the sampling variability of the disease outcome. The following results are given by Greenland and Brenner (1993). Imperfect disease specificity (i.e. false positive rate) induces no bias in the risk difference estimate, but this estimate is biased toward the null value by imperfect sensitivity. With disease misclassification, due either to imperfect sensitivity or to specificity, the linear regression estimate of the risk ratio will be biased towards the null hypothesis (i.e. equal risks in the compared populations).

Another important issue in the study of ecological data concerns the sensitivity of these analyses to model mis-specification. The mathematical form of a model depends on many issues. The ecological relation in a particular group embodies the group means for the disease rate, exposure variate and covariate level across exposure–covariate strata. In general, this relation does not assume the functional form used in individual level studies (Richardson *et al.*, 1987). This discrepancy may cause only little bias when the expected effects are small. Nevertheless, ecological summary rate ratios can be very sensitive to the chosen model form (Greenland, 1992). In contrast, the individual-level effect summaries of rate ratios appear insensitive to the choice of model structure (Maldonado and Greenland, 1993).

The choice of the regression model also has implications on the epidemiological inferences. In linear additive models, the estimated values of the disease rate parameters must be restricted so that they predict positive rates. To overcome the possible problem of extrapolation to negative values for rates, a logarithmic transformation of the rate parameter can be used. This transformation implies, however, that all continuous terms in the multiplicative model assume an exponential relation to one another. It is also difficult to find foolproof programmes to fit these models. Fortunately, the advances in the generalised linear (McCullagh and Nelder, 1983) and additive (Hastie and Tibshirani, 1990) models have opened new possibilities for more versatile modelling of ecological data.

Failure to identify, measure, or control important covariates of the dose–response relation is known as confounding. This problem is shared by cohort and case-referent study designs as well as all types of ecological design, but the problem is more perplexing in ecological studies than in individual level studies. This is true because ecological bias can be produced by other factors, such as effect modifiers acting independently of the confounders or tangling with their effects. Thus, the conditions for no confounding in ecological studies are logically independent of the conditions that guarantee no confounding in individual level studies. If the latter conditions are mistakenly applied in ecological studies, it can lead to omission of important covariates from the analysis. As Greenland (1992) argues, a covariate may be ignored at the individual level but not on the ecological level, or vice versa. There will be no ecological association of exposure distributions with disease outcome rates across groups if there are no exposure effects on either disease risk or the distributions of other risk factors by group (Greenland and Robins, 1994). This condition occurs even if, within each group, exposure levels are associated with the other risk factors.

If existing databases are used, the extent of those information sources imposes certain limitations. The use of routinely collected health and environmental data, by necessity, restricts confounder control possibilities to those covariates that have been measured. These variates usually do not include all the relevant covariates for the studied exposure–effect relation. Moreover, most covariates used in ecological regressions are either surrogates or crude measures of the true confounder. The problem is compounded by the need to measure the multivariate (joint) distribution of the confounders; univariate distributions of the covariates or a confounder score may not suffice to achieve full control of confounding.

The effect of exposure on disease outcome can vary according to the level of a covariate; this is termed "effect modification". These covariates introduce statistical interaction with the exposure variate. Any individual level study that records exposure and covariate can analyse their interaction by including their product in the regression model. In an individual level cohort study, omission of the product term from the regression model induces no bias (Greenland, 1992). In stark contrast, omission of the group mean of the product term from the ecological regression model can lead to severe bias when this model is correct. The summary estimates of rate differences and rate ratios may even lie outside the range of their true covariate-specific values (Greenland and Morgenstern, 1989; Greenland and Robins, 1994). Ecological bias caused by effect modification across areas can occur even when the number of areas is very large and there are no across-area or within-area associations of exposure with the covariate, and thus there is no confounding (Greenland and Morgenstern, 1991). In contrast, effect modification cannot by itself produce bias in individual level analyses (Miettinen, 1985).

There is need for standardisation in ecological studies in epidemiology for variates where the distribution is not constant across population groups. This is important because published disease rates are invariably age-standardised, whereas published exposure rates are seldom standardised. In regression analysis of ecological data, the covariates need to be mutually standardised using the same standard distribution as used for the disease outcome. If a different standard (or no standardisation) is applied to the covariates, the inclusion of the confounding variates in the regression model may even aggravate the bias. Similarly, if only the outcome rate is standardised, the bias may consequently increase. In addition, the exposure variate must be standardised using the same standard distribution; otherwise standardisation bias may result (Greenland and Morgenstern, 1991).

In the study of chronic diseases using ecological studies it is important to consider the induction period. Ecological estimates of exposure may not correspond to the aetiologically relevant period for diseases with long induction times. If the exposure coincides more with that of the health outcome than with the relevant aetiological time, the relation is cross-sectional. In a longitudinal relation, exposures at some previous time(s) are considered. If the exposure level is stable over time, the distinction between the two relations is debatable. If it is very unstable, the distinction is critically important. Thus there is need to focus on the most relevant time span of the aetiological period.

Dynamic populations are characterised by in-migration and out-migration which may render the areal exposure estimates at earlier times poor proxies for the actual levels experienced by the study populations providing the disease rates (Greenland, 1992). One way of tackling the problem of migration would appear to be to restrict ecological studies to more homogeneous populations. However, this restriction may lead to a vicious circle, because it generally means that the groups studied are smaller and, therefore, more easily subject to differential representation of the exposed or unexposed domains in the study base. Geographically static populations may also be unrepresentative of the wider population.

Failure to distinguish between the scientific object of epidemiological research and the actual object of an empirical study may lead to incorrect inferences. In aetiological research, the object of inference is the same for both ecological and individual level studies. Although in individual level studies the target effects are at the same level as the units of statistical analysis, the ecological analysis is coarser. Consequently, an ecological study can yield biased results for individual level effects whilst still being unbiased for ecological effects (Greenland and Morgenstern, 1989).

6.2.4   Strategies for minimisation of bias

In the preceding section, the sources and directions of various biases were considered. The primary strategy for the prevention of bias in ecological studies, as in epidemiological studies in general, must be the design of a valid study. If bias persists, there are limited statistical methods available for reducing bias in the analysis phase, although these will rarely eliminate bias entirely. They include:

- Multilevel modelling of the exposure–effect relation.
- Coping with model assumptions.

- Influence analysis.
- Sensitivity analysis.
- Use of robust procedures.
- Empirical Bayes methods.
- Correcting for nondifferential misclassification.
- Controlling for confounding.
- Modelling for nonlinearity and nonadditivity of effects.
- Adjusting for variates using comparable standardisation.

Because of the developments in epidemiological study design and the progress of analytical multilevel modelling techniques (also called hierarchical regression (Greenland, 1998)) it may be possible to alleviate the problems with inference based on grouped data by obtaining individual level data in samples of selected groups. Multilevel modelling allows for the simultaneous analysis of individuals and their ecologies. This approach examines the circumstances of individuals at one level and, simultaneously, the contexts or ecologies in which they are located at another level. The result is a strategy for the analysis and adjustment of aggregation effects in a regression analysis. In addition, the modelling provides a way of removing the bias due to the grouping variates if additional information about the individual-level covariances between the grouping variates is available. Pearce (1999), however, argued that *"Epidemiologists need to learn to think in a multi-level way, rather than just adding a multi-level modelling into their analytical toolkit"*.

Most ecological studies are descriptive, and as such can be classified as exploratory data analysis. Influence analysis extends data description to exploration of the sensitivity of data summaries or exposure effects. It can, for example, be especially informative in ecological studies with small groups to examine the impact of excluding some observations which stand out as statistical outliers.

Because ecological estimates are sensitive to biases, it is necessary to fit multiple regression models in an ecological analysis, particularly when the true form of the model is unknown. Unfortunately, the parametric model for risk of disease is almost always mis-specified in practice. Greenland (1979) points out in the context of discussing the limitations of the logistic analysis of epidemiological data, that *"... as with all statistical models, there is a danger that the ease of application of the model will lead to the inadvertent exclusion from consideration of other, possibly more appropriate models for disease risk."* The choice of the model has consequences for the effect estimates because of the frequent need in risk assessment for extrapolation to zero. In a sensitivity analysis, the various model forms are varied to check the invariance of the results under new distributional assumptions.

An answer to the uncertainty about model specification is to employ robust methods (i.e. methods that function better than usual methods when the assumptions underlying the usual methods are violated). Random effects models can be used when the distribution of observations under the usual probability model show overdispersion, i.e. the mean exposure levels can exhibit extra-normal variation, or the disease occurrences can exhibit extra-Poisson variation. Another proposed remedy is to use empirical Bayes methods which can successfully deal with several epidemiological problems, such as disease mapping, smoothing of unstable rates and screening of multiple associations (Greenland, 1994).

Greenland and Brenner (1993) have provided a general method of adjustment for nondifferential misclassification of a binary exposure variate (e.g. smoker or non-smoker) and disease outcome in ecological regression analysis. They derived simple correction formulae for the ecological regression estimates of risk difference and risk ratio. This method uses the concepts of sensitivity and specificity of exposure and disease measurement. The method can be applied specifically for exposure (Brenner et al., 1992b), disease outcome (Greenland and Brenner, 1993) and confounders (Brenner et al., 1992a).

The degree of confounding, and hence the strength of an effect, cannot be directly measured from the observed data; it should be evaluated against background disease risk, knowledge of subject-matter, logical argument, evidence from previous studies, and the particulars of the empirical setting in which the study is being conducted (Nurminen, 1997a). In general, the control of confounding in an ecological study is more demanding than in an individual level study because the measurement process for confounders is much more complicated. As in an individual level study, the ecological approach entails the problem that the crude measurement or approximation of a confounder may be inadequate for achieving full control. Moreover, for potential confounders, such as diet, smoking and other lifestyle factors, multiple summaries of the joint distributions are needed for effective control. Unfortunately, the within-group joint distribution of the covariates is rarely available in ecological studies. In particular, the marginal summaries that may be available may prove to be too crude to provide effective control. If, however, confounder information is available in the disease registration system or is estimated from sample surveys, then the analysis can be improved by stratification by the covariate (Brenner et al., 1992a).

The problem of confounding is aggravated if nonlinearity or nonadditivity of effects by the covariate are present. Ecological covariate summaries can be inadequate to detect and control confounding by a covariate with non-linear effects, and also when the effects are not additive (i.e. in the presence of effect modification). In addition, the covariate terms in the regression function must be adjusted to the same distribution as the disease and exposure using comparable standardisation.

6.2.5   Designing, analysing and evaluating ecological studies

Ecological studies are based on a distinct methodological approach which sets them apart from individual level epidemiological studies. A number of specific factors thus have to be considered in either designing or analysing ecological studies, or in critically evaluating the end-results of such studies (Morgenstern, 1982; Greenland, 1992):

- Ecological studies are much more sensitive to bias from model mis-specification than are results from individual level studies. For example, deviations from linearity in the underlying individual level regressions can lead to inability to control confounding in ecological studies, even if no misclassification is present.
- Conditions for a covariate not to be a confounder differ in individual level and ecological analyses. Thus, a covariate may be negligible at the individual level but not at the ecological level, or vice versa.
- In contrast to individual level studies, independent and nondifferential misclassification of a dichotomous exposure variate usually leads to bias away from the null hypothesis in ecological studies.
- Failure mutually to standardise disease, exposure and covariate data for other confounders (not included in the regression model) can lead to bias.
- There is no ecological method available to identify or measure ecological bias.
- In the design of an ecological study, it is important:
  - to select areas with populations that are as homogeneously exposed as possible (i.e. minimise within-area exposure variation) by sampling smaller units for the analysis;
  - to select populations which represent different extremes of exposure distribution (i.e. maximise between-area exposure variances);
  - to select populations which are comparable with respect to covariate distributions; and
  - to supplement, whenever feasible, approximate aggregated data with accurate data at the individual level in a hybrid epidemiological analysis.
- In the analysis of ecological data it is important:
  - to use weighted regression, instead of correlation, with weights proportional to the amount of information contained in each group;
  - to include in the regression model all variates that are thought to be related to the grouping process;
  - to examine multiple regression models with different and flexible structural forms beyond the standard linear form, such as exponential and product-term models, and nonparametric, smoothed curves;

- to test the basic assumptions in the model (i.e. the robustness of estimation techniques);
- to consider the ecological implication of specifying different individual level forms of model;
- to conduct an influence analysis by examining the effect of deleting from the analysis various areas with unusual outcome, exposure or covariate combinations;
- to conduct a sensitivity analysis of ecological estimates to misclassification;
- to take into account latency and induction periods separating causes and effects (e.g. consider the relevant exposures);
- to consider the effect of migration on areal exposure estimates; and
- to accompany ecological analysis by thorough consideration of biases unique to such an analysis, and to biases common to all epidemiological studies.

6.3   Time series analysis

Time series analysis (TSA) is a well established technique in statistics. It was developed to a large extent for econometric applications, but has since been adopted in a wide range of disciplines. Time series analysis is typically used to investigate patterns in series of observations, as a basis for identifying and quantifying causal relations. In environmental epidemiology, it is often applied to long sequential observations, such as mortality statistics, data from morbidity registers (e.g. cancer registers, hospital discharge registers), or results from repeated health surveys. In the case of HEADLAMP studies, TSA offers a valuable means of examining and comparing trends in environmental conditions and health effects in a study area, or of assessing the effects of policy actions on environmental conditions and health outcome. Even though there does not seem to be anything unique about time that is fundamentally distinct from any other covariate (e.g. space) that defines the sampling units in an ecological study, the importance of time series warrants their separate consideration.

With simple data sets, TSA is a relatively straightforward method, and is supported by most well-equipped statistical packages. Where temporal patterns are complex (and thus where relatively complex models need to be used to describe the time series), however, it can be computationally and statistically demanding, and can pose severe problems for both implementation and interpretation. In recent years, it has been extensively applied in studies of air pollution and health, and thus efforts have been made to formalise and standardise the techniques used (e.g. Katsouyanni *et al.*, 1997). Moreover, as temporal data series are extended and improved, the opportunities for TSA will inevitably increase. Continued interest in the use of TSA may thus be expected.

Time series analysis looks at the relation between observations recorded at consecutive, usually equally spaced, discrete time points. While TSA is also a regression method, it predicts the health outcome not from independent covariates, but from values of the outcome at previous points in time. The minimal requirements are the abilities to plot the temporal series; to derive new series (e.g. differenced series or smoothed series) and to plot these; to examine scatter plots of time-lagged values; to compute serial correlation periodograms; and to display these graphically. Current developments in graphical computing techniques for studying multidimensional relations will be valuable for TSA. Tools for statistical computing are especially important when the data sets used are large.

Three basic approaches to TSA exist, namely:

- Poisson autoregression analysis using generalised estimating equations (GEE).
- Markov regression models using quasi-likelihood estimation (QLE).
- Poisson risk function model for time-stratified data using maximum-likelihood estimation (MLE).

It is beyond the scope and depth of this chapter to present the details of the statistics involved and therefore the following sections outline, in general terms, how these models are applied in TSA.

6.3.1   Regression models for time series analysis

The time series model can be understood as a subclass of the generalised linear model (McCullagh and Nelder, 1983) in which the exposure effects are multiplicative, the distribution of the errors is Poisson, and the link function is the (natural) logarithm. Thus, the model can be expressed as:

$$\log\left[E(y_t)\right] = x_t^{'}\beta$$

where: y_t is the count of observed outcomes at time t,

$E(y_t)$ denotes the expected count,

x_t is the (column) vector of covariates, and

β is the vector of regression parameters.

In this model, β represents the effects of the covariates on the outcomes.

To account for the possibility of overdispersion and autocorrelation, the covariance matrix for the health outcomes on the units of observation is assumed to have a special form; the regression parameters are then estimated by the GEE (Liang and Zeger, 1986). Generalised estimating equations are used because the form of the joint distribution of the time-dependent measurements is so complex as to be intractable; i.e. it cannot provide useful and interpretable information.

Overdispersion in Poisson counts can arise for at least two reasons. First, the risk of an adverse outcome occurring to an individual may not be equal for all individuals, but may depend on previous events that happened to that individual; i.e. it varies over time. The second reason is that the risk may remain constant over time but not necessarily be equal for all individuals.

Markov models can also be applied for regression analysis of time series data (Zeger and Qaqish, 1988). As serial observations are unlikely to be independent, in the Markov models the expected response at a given time depends not only on the associated exposure variates and covariates but explicitly also on health outcomes at previous times. The regression coefficients can be estimated using the QLE approach (McCullagh and Nelder, 1983). Quasi-likelihood estimation allows the regression relation to be estimated without full knowledge of the error distribution of the response variate.

There is a fundamental distinction between the GEE approach, which is a "pure" regression model with autocorrelated errors, and the QLE approach, which is a mixed regression–autoregression model. Although these models may be considered as alternatives, in general the regression coefficients in the two models are different because, as in any regression equation, the interpretation of a parameter depends on what other variates are included in the model. It is also normally inappropriate to assume that the error in the exposure variates is negligible. When data on measurement errors is lacking, estimates should be obtained from independent survey samples. An advantage of the QLE approach over the GEE approach is that competing models can be compared directly with each other using a deviance statistic.

A particularly problematic aspect of studying temporal relations arises when there are sharp peaks present of similar frequency in both response and exposure series. For example, in an epidemiological study of daily death rate and meteorological variates, seasonal fluctuations are likely to be present in all data sets. In other instances, troughs or long-term trends may be present. Time series analysis deals with this by studying the regressions separately in different seasons or periods. More commonly, TSA considers differences in adjacent time points, and the exposure–effect relation between them is modelled. An assessment is then made of whether any model so fitted accounts for all the relation present.

Previous use of Poisson autoregression analysis models has generally been based on the assumption that the series is time-dependent. Nevertheless, it is not clear either that the GEE approach or the QLE approach has advantages over simpler model building procedures sufficient to compensate for their greater statistical complexity. All the autoregressive methods involve complex and computer-intensive estimation procedures. A much simpler way of dealing with temporal data may be to adopt the working assumption that repeated observations from a unit are time-independent of one another. It is possible then to proceed by dividing the study data into subgroups (strata) and fitting a Poisson risk function model to the time-stratified series. In this approach, the assumption of constant risk (or rate) ratio can be alleviated by including time-dependent covariates in the linear predictor. Computational demands can be reduced by using the MLE methods available with existing software. Kuhn *et al.* (1994), for example, used this method for TSA and found that it compared favourably with the GEE approach.

6.3.2   Application of Poisson regression for time series

Given the potential use of Poisson regression to quantify time trends, it is worthwhile to consider some assumptions of the MLE method that may appear to be violated in the case of time series data. Simple Poisson modelling requires that outcomes are independent. On first thought, it would seem untenable to assume that health outcomes occurring over time meet this requirement. For example, the effects of social and environmental conditions are likely to persist at least in the short run. Poisson regression also assumes that the population subgroups are homogeneous with respect to the risk of adverse health outcome. This is another questionable assumption because the occurrence of, for example, asthmatic or cardiac attacks do not occur at random but have predictable precursors and known patterns of risk.

Two points need to be emphasised here. The first is that the ordinary Poisson regression model requires that the study population meets the criteria of no overdispersion and heteroscedasticity conditional on the covariates. One effective way of removing overdispersion is to transform the data to a square root scale; this stabilises the variance of the observed counts. The inclusion of time-dependent covariates may well result in conditional independence and help to define strata of homogeneous risk. Secondly, the Poisson regression allows the analysis of aggregate data to be comparable with the analytical methods used in cohort and case–cohort (or case–base) studies (Nurminen, 1992). Thus, although autoregressive time series analysis has been promoted as the preferred method in analyses of sequential observations over long periods of time, it may equally be argued that Poisson regression provides a simple and viable alternative for time stratified data.

The minimum requirements for epidemiological studies using time–trend ecological studies are basically the same as those required of a valid and precise epidemiological study in general. Inadequacies in the database and the sheer complexity of interactions among relevant variates both add to the problem of inferring the exposure–effect relation between pollution and health. The HEADLAMP approach is based on the idea of relying on established associations between environment and health, and should be applied on a local or national level to infer about excess risks. Thus it is not necessary to speculate on the kind of biological mechanism behind findings of such an associative nature. However, even within this limited framework, at least part of the difficulty stems from methodology. An appropriate application of statistical methods for regression must necessarily cope adequately with the time series characteristics of pollution and health variates.

In general, the representation of confounders in the regression model should be guided by the concern for thoroughness of control, with the reservation that the efficiency of the study not jeopardised by the inclusion of covariates of the exposures which are not risk factors of the disease outcome. A major drawback of a time series design is the possible presence of unmeasured confounders. However, the time–trend study of short-term effects that uses long series of small units (days) often downplays such errors. An important feature of such studies is that the population followed up serves as its own control over time, and thus possible confounders can only be factors varying according to small time units (from day to day). Such factors might be meteorological and chronological factors; fortunately these are usually accurately measured and easily recorded.

6.4   Quantitative risk assessment

Increasingly, authorities at the local, national and international level are faced with difficult decisions which involve weighing the social and economic benefits of technology against the health and environmental costs involved (McMichael, 1989). If these decisions are to be made on an informed basis, they require that health effects can be quantified. As a result, some form of quantitative risk assessment (QRA) is necessary for regulatory purposes. Moreover, because the results of such assessments are often presented as a single number (for example, excess number of exposed disease cases), they give the appearance of scientific certainty and simplicity, both of which make the methods appealing to decision-makers. In practice, however, the ability to quantify the health effects of development is often limited, and valid methods of QRA are both uncertain and complex. For example, the methods are highly dependent on a series of assumptions and subjective choices which can have critical effects on the resulting risk estimates. Considerable care is therefore necessary in using and interpreting results of QRA. For a review of methodological issues in epidemiological risk assessment, see Nurminen *et al.* (1999).

Quantitative risk assessment can be defined as the application of a statistical relation between exposure and the associated health outcome to assess either the health risk to a population or the exposure level associated with a given risk. Thus, two main types of QRA can be distinguished:

- Risk analysis, which involves computation of the risk corresponding to a given level of exposure or dose; for example expressed in terms of excess risk or the number of extra disease cases.

- Hazard analysis, which involves calculation of the exposure or dose corresponding to a given level of risk; for example the exposures estimated to cause adverse health outcomes in a certain percentage of exposed subjects.

Risk analysis may be also applied at two different scales. Individual risk refers to the probability that an individual will develop a disease as a result of exposure in a specified time period. Population risk refers to the expected number of cases of disease attributable to exposure in the population under study in a specified time period. These two measures may have different regulatory implications: the regulatory authorities may wish to evaluate either the risk to individuals who are exceptionally highly exposed or the risk to a large population for which the average exposure could be much lower.

6.4.1   Uses and uncertainties of quantitative risk assessment

Risk analysis is not a true linkage method in the sense that local health data are not used. Instead, it uses a predefined association between exposure and health outcome to determine the risk to an exposed population. The relation between exposure and health is usually derived from independent studies, either within the study area or, more commonly, elsewhere. The particular advantage of risk analysis is thus that it can be applied in areas where insufficient health outcome data are collected to allow the relationship between exposure and health to be determined locally. By the same token, QRA methods are the least resource-intensive, the easiest and the fastest to use of all the methods considered here. The success of the risk assessment process, however, depends on a number of issues, such as the choice of the risk prediction models and the adequacy of exposure assessment. All of these are subject to large uncertainties, although the exact form and magnitude of these problems vary depending on the particular context and purpose of the analysis.

One of the most important difficulties in QRA lies in obtaining reliable estimates of the exposure–response relation. Results from epidemiological studies of one population cannot always be directly applied to others, due to differences in the range of exposures involved, in the methods of exposure estimation used, in the socio-economic contexts in which exposure occurs and in the baseline status of the populations concerned. A relation for exposure to air pollution derived from a developed country or city, for example, is likely to underestimate the risks in developing countries, where the baseline health status is poorer (Ostro, 1994). Similarly, differences in the way in which exposure or health outcome are defined or measured in different areas (e.g. in the design of the pollution monitoring network, the specific definition of the pollutants measured, or in diagnosis) may make it difficult to transfer relations from one area to another.

Particular care is also needed where the health outcome of concern is potentially related to more than one exposure. Both particulate matter and

SO_2, for example, are known to contribute to respiratory diseases. In many areas, levels of the two pollutants are also highly correlated. When modelling the contribution of both, only one variable will remain statistically significant — the effect of the second will be subsumed within the first. When modelled separately, on the other hand, they may both show significant associations with health outcome. Summing these separate estimates of the effects clearly exaggerates the estimated effect (e.g. the likely number of cases). Ideally, therefore, some measure of the combined effect should be obtained, by adjusting for the effect of the second exposure. In practice, this is often difficult, and in these cases a more conservative approach is to use only one measure of exposure — perhaps the one with the more complete data set.

A further source of uncertainty in QRA is the presence of population heterogeneity. In environmental health linkage, risk factor data are usually collected and presented at high levels of aggregation. Aggregated risk estimates of this type can only be extrapolated back to the individual level if the population concerned is homogeneous. In reality, homogeneity within any population rarely if ever exists. Unrecognised risk factors may be expected to subject different people to different background disease risks. As a result, individual risks may differ substantially from those implied by the aggregated data. Usually, variance estimators tend to be invalid when risks are heterogeneous. In undertaking a risk analysis, therefore, it is necessary always to check for hidden heterogeneity before presenting aggregate population statistics. If heterogeneity is discovered, then population risk estimates based on the aggregate data may be misleading. The populations should either be subdivided into more homogeneous subpopulations, or the statistics should be presented with due cautions for interpretation.

6.4.2 Presentation and interpretation of results of risk assessments

The results from any risk assessment clearly need to be communicated to the decision-maker in an appropriate form. This implies that the results are both clearly presented, yet also suitably qualified with regard to their reliability. The interpretation of the results, both by the risk assessor and the risk manager, and later by the governmental and non-governmental organisations as well as the general public, may be critically dependent of the methods used to present the results. This is especially crucial in linking environment and health data because the decision-makers may not be well versed in the specialised statistical methods used. Moreover, there is the need to present the linkage results in such terms that they can be transformed easily to inputs for a societal or an individual cost-benefit analysis, or disseminated to other stakeholders (e.g. the public).

At present, there are no standardised procedures for analysing and presenting results from environmental and health linkage. To a large extent, this reflects the many different methods used to analyse the data, and the inherent differences in the data themselves. As a result, a standardised approach for the linkage of environmental health data is often neither feasible nor necessary. It may not be feasible because of irresolvable differences in the data or methods available and it may be unnecessary because the study concerned does not involve comparisons across different areas or periods.

Standardisation of methods is nevertheless beneficial insofar as it facilitates comparability. The diversity of analytical techniques so far applied in time–trend studies of air pollution and health, for example, has tended to hinder direct comparisons of the results, and has made it difficult to derive general estimates of exposure–effect relations (e.g. from a meta-analysis). Lack of standardisation also makes it difficult to verify the results of individual studies (e.g. by comparison with studies elsewhere) and reduces the opportunities to reuse the data at a later date. Standardisation thus offers the possibility of obtaining added value from the data, and thereby of improving the cost-effectiveness of data collection. One of the rare attempts to establish standardised procedures for time series analysis was the EU funded APHEA project (short-term effects of Air Pollution on Health: a European Approach). This approach developed a standardised methodology to analyse data from 15 cities, representing a range of social, environmental and air pollution conditions across ten countries (Katsouyanni, 1997).

It is beyond the scope of this chapter to discuss how best to present the results of statistical analyses because they are well covered in many text-books (e.g. Gore and Altman, 1982). Similarly, the technicalities in quantifying human health risks are not considered here because they have been described in books on risk assessment (e.g. Cox and Ricci, 1989). It is, however, useful to examine some of the general issues involved in the presentation of the results of linkage studies, as a basis for better informing the decision-maker.

The result of most interest to the health agency or risk manager in arriving at a decision, is usually the quantitative estimate of exposure effect on health risk. It is this effect estimate which provides the platform for subsequent policy action. Two quantitative measures of effect are widely used:

- Change in individual risk, i.e. the increased or reduced likelihood of an individual experiencing a specified health effect due to a change in exposure level.
- Disease burden, i.e. the number of excess cases of the specified health effect ("body count").

Table 6.1, for example, shows the average working life risk of lung cancer for an individual exposed to silica, while Table 6.2 shows the excess numbers of lung cancer in the exposed population for both the currently prevailing exposure levels and for the lower control limits. To provide some

Table 6.1   Average lifetime risk of lung cancer for silica-exposed men employed from age 20 to 60 years

Exposure level	Estimated risk (%)	95% confidence interval
Current	1.87	0.21–31.3
≤0.2 mg m^{-3}	1.34	0.20–3.98
≤0.1 mg m^{-3}	0.83	0.15–1.92

Source: Leigh *et al.*, 1997

perspective, the results of risk assessment are often expressed as hypothetical changes in risks. Thus, a risk-analyst might interpret the results of Table 6.2 as follows: introduction of, and adherence to, an exposure standard of 0.2 mg m^{-3} would produce a 22 per cent reduction in the excess number of lung cancer cases. Alternatively, if the exposure standard was set at 0.1 mg m^{-3}, a 46 per cent reduction would be predicted.

The methods used for risk estimation inevitably give only approximate projections of risk, because they usually involve a myriad of assumptions, which cannot easily be verified. The presentation of simple point estimates of the expected risks and excess numbers thus tends to give a misleading impression of precision. Instead, it is important to provide clear information on both the assumptions and limitations involved. Cox and Ricci (1989), for example, suggest the following guidelines for the presentation of risk estimates:

- Risks should be presented in a sufficiently disaggregated form (showing risks for different subgroups) so that key uncertainties and heterogeneities are not lost in the aggregation.
- Confidence bands around the predictions of statistical models are useful, but uncertainties about the assumptions of the model itself should also be presented.
- Both individual risks and population risks should be presented, so that the equity of the distribution of individual risks in the population can be taken into account.
- Any uncertainties, heterogeneities, or correlations across individual risks should be identified.
- Sensitivity analyses should be used to assess the effects on estimates of the key assumptions involved.

Linking environmental exposures to health outcomes is frequently achieved through the use of a regression model, for example a multiple logistic regression. Whatever method is used, presentation of results after allowance for covariates should be in a form similar to that which would be used if no covariates were included in the risk function. Merely quoting the

Table 6.2   Excess lung cancer cases in a dynamic working population of 136,400 men exposed to silica in a 40-year follow-up period

Exposure level	Estimated number	95% confidence interval
Current	1,960	300–7,760
≤ 0.2 mg m^{-3}	1,524	270–3,530
≤ 0.1 mg m^{-3}	1,058	210–2,200

Source: Leigh *et al.*, 1997

coefficients from the logistic model does not achieve this and is in any case artificial, because the logit transformation would not be necessary if there had been only the one risk factor of interest, and no covariates. This does not mean that the risk–odds ratio would not be useful as an auxiliary parameter in risk modelling. The analyst should, however, also provide more informative measures of exposure effect, such as the absolute excess risk (risk difference) or the relative excess risk (risk ratio minus one) (see Nurminen, 1995b).

A minor, yet more than cosmetic, point in presentation of results from QRA is the number of significant figures. In this context, the inherent precision of the results needs to be acknowledged. It is not sensible, for example, to give a result as "49.35 expected disease cases per year" when the probable range is from 10 to 200. It might even be better not to give a single point estimate, but only to indicate the approximate confidence bounds. In presenting the results of a meta-analysis, the overall mean value can be shown along with the ranges for the lower and upper confidence limits.

Quantitative risk assessment frequently presents information in terms of probability measures. Probability distributions can be difficult for a non-specialist to interpret. Although a plot of cumulative incidence rate (estimates of risk) allows one to read the median (and the percentiles of the distribution), the mean value cannot be determined from the plot. To avoid misinterpretations, therefore, it is important to present a plot of the cumulative distribution together with a graph of the incidence density curve, using the same horizontal scale, and to show also the mean risk on both curves (Ibrekk and Morgan, 1987).

To be of use for health policy making, epidemiological data often need to be interpreted. Epidemiologists are mostly concerned with the increased incidence associated with exposure to a risk factor, whereas policy-makers are more interested in the reduction of risk after the cessation of exposure. The importance of a risk factor for the incidence of a disease in a population is usually expressed as the aetiological fraction, that is the proportion of the total incidence of the disease that can be attributed to that risk factor in the population (Miettinen, 1985). This indicates the maximum proportion of incidence that could be prevented by the elimination of that risk factor within the population.

In practice, prevention measures are rarely able to eliminate, but merely reduce, the prevalence of an environmental risk factor. As a result, a more useful measure is the potential impact fraction (Morgenstern and Bursic, 1982). This indicates the incidence that is avoided by a preventative intervention as a proportion of the incidence that would have occurred in that population without intervention. The potential impact fraction can be calculated when the prevalences of exposure to a risk factor in the population and the corresponding incidence density ratios or risk ratios are known.

In the traditional epidemiological literature, the term potential impact fraction is often used to imply an immediate removal of excess risk after termination of exposure. In reality, this risk reduction may take many years to achieve, due to the lag effects involved. Ideally, therefore, estimates of effect should incorporate a time dimension. For this purpose, a methodology based on the preventative impact fraction has been developed (Gunning-Schepers, 1989). This comprises a computer simulation model, PREVENT (Gunning-Schepers et al., 1993), that can estimate the health benefits for a population of changes in risk factor prevalence. Results are presented in graphical or tabular form and include the intermediate output variates (aetiological fraction, trend impact fraction, and potential impact fraction) and the final output variates (disease-specific mortality, total mortality, disease-specific mortality difference, potential years of life gained, actual years of life gained, survival curves, and life expectancy at birth).

A preventative intervention programme is often difficult to "sell" politically because its effects take so long to become apparent. Indeed, in many cases, the effects are not expressed as real reductions in risk because of the demographic changes in the target population over time. This does not mean that prevention will have no beneficial effect. It does mean, however, that in order to see the effects it is important to show what would happen without the preventative intervention, and not merely to compare predicted effects with the current level of mortality. The potential use of simulation models, such as PREVENT, in this respect lies in their ability to provide more precise quantification of effect estimates over time, and to take account of multiple risk factors and possible effects of demographic changes on the effects of intervention (Gunning-Schepers et al., 1993).

Although risk estimates produced by risk analysis have been used traditionally as the justifiable basis for regulating risks, the public's perception of risk is much broader than the "body counts" on which the quantitative risk assessments have focused. The public frequently misperceive risks because of the biases in the information to which they are exposed (e.g. the news media, government reports and industry reports). The public also perceives risk in a much wider context than that used in environmental epidemiology, i.e. perceptions reflect dread of the unknown, social and political impact, outrage and stigma. This difference in risk perception calls for two-way communication between the risk-analysts, risk-managers and other policy-makers, on the one hand, and the general public on the other (Morris, 1990). Useful guidelines and suggestions on how to communicate results of QRA to the public have been published by the US Environment Protection Agency (Covello and Allen, 1988). These list "cardinal rules" for effective risk communication. In addition, a useful guide designed for industrial plant managers is available that describes the technical information to be presented and provides guidelines for explaining risk-related numbers and risk comparisons (Covello *et al.,* 1988).

6.5   Conclusions

The linkage of environmental and health data (or either of these with covariate data such as socio-economic or demographic information) is a vital part of the HEADLAMP approach. Unlike in traditional epidemiology, its aim is not to seek new environmental health relations or confirm hypotheses; rather it is to use existing knowledge on such relations to help inform management and policy decisions, and to raise awareness about the associations between environment and health. The methods are thus used essentially as a means of describing and monitoring the relations between environment and health, and to help assess and demonstrate the existing risks to the population concerned.

Any such data linkage must, nevertheless, be undertaken with care, because the relations between environment and health (whether expressed geographically or in terms of time trends) are often complex and fraught by uncertainties. Without an understanding of these complexities, it is all too easy to misinterpret the data. On the one hand, this may lead to complacency and lack of action, if risks are not correctly identified; on the other it may cause unnecessary anxiety and fear, if non-existent risks are inferred. It is important to recognise that these dangers may be created simply by presenting environment and health data together, because it is human nature to search for associations. Since most observers will be unaware of the complexities and subtleties of the data, misinterpretation is almost inevitable. Data linkage thus needs to be recognised as a powerful but potentially treacherous tool. It is incumbent on the analyst, therefore, to ensure not only that environment and health linkage is conducted rigorously, but also that the results are presented and explained clearly and unambiguously.

6.6   References

Brenner, H., Greenland, S. and Savitz, D.A. 1992a The effects of nondifferential confounder misclassification in ecologic studies. *Epidemiology* **3**, 456–9.

Brenner, H., Savitz, D.A., Jöckel, K.-H. and Greenland, S. 1992b Effect of nondifferential exposure misclassification in ecologic studies. *American Journal of Epidemiology,* **135**, 85–95.

Briggs, D.J., and Elliott, P. 1995 The use of geographical information systems in studies on environment and health. *World Health Statistics Quarterly,* **48**, 85–94.

Cochran, W.G. 1977 *Sampling Techniques.* Third edition. Wiley, New York.

Covello, V. and Allen, F. 1988 *Seven Cardinal Rules for Risk Communication.* United States Environmental Protection Agency, Washington, D.C.

Covello, V.T., Sandman, P.M. and Slovic, P. 1988 *Risk Communication, Risk Statistics, and Risk Comparisons: A Manual for Plant Managers.* Chemical Manufacturers Association, Washington, D.C.

Cox Jr., L.A. and Ricci, P.F. 1989 Risk, uncertainty, and causation: Quantifying human health risks. In: D.J. Paustenbach [Ed.] *The Risk Assessment of Environmental and Human Health Hazards: A Textbook of Case Studies.* Wiley, New York, 125–56.

Elliott, P., Cuzick, J., English, D. and Stern, R. 1992 [Eds] *Geographical and Environmental Epidemiology. Methods for Small-Area Studies.* Oxford University Press, Oxford.

Fuller, W.A. 1987 *Measurement Error Models.* Wiley, New York.

Gore, S.M. and Altman, D.G. 1982 *Statistics in Practice.* British Medical Association, London.

Greenland, S. 1979 Limitations of the logistic analysis of epidemiological data. *American Journal of Epidemiology,* **110**, 693–8.

Greenland, S. 1992 Divergent biases in ecologic and individual-level studies. *Statistics in Medicine,* **11**, 1209–23.

Greenland, S. 1994 Hierarchical regression for epidemiological analyses of multiple exposures. *Environmental Health Perspectives,* **102**(Suppl. 8), 33–9.

Greenland, S. 1998 Introduction to regression modelling. In: K.J. Rothman and S. Greenland [Eds] *Modern Epidemiology.* Second edition. Lippincott-Raven Publishers, Philadelphia, 401–32.

Greenland, S. and Brenner, H 1993 Correcting for non-differential misclassification in ecologic analyses. *Applied Statistics*, **42**, 117–26.

Greenland, S. and Morgenstern, H. 1989 Ecological bias, confounding, and effect modification. *International Journal of Epidemiology,* **18**, 269–74.

Greenland, S and Morgenstern, H. 1991 Neither within-region nor cross-sectional independence of exposure and covariates prevents ecological bias (letter). *International Journal of Epidemiology,* **20**, 816–7.

Greenland, S. and Robins, J. 1994 Ecologic studies: biases, misconceptions, and counterexamples. *American Journal of Epidemiology,* **139**, 747–59.

Gunning-Schepers, L.J. 1989 The health benefits of prevention, a simulation approach. *Health Policy,* **12**, 1–256.

Gunning-Schepers, L.J., Barendregt, J.J.M. and van der Maas, P.J. 1993 PREVENT, a model to estimate health benefits of prevention. In: *Tools for Health Futures Research.* World Health Organization, Geneva, 7–21.

Hales, S., Lewis, S., Slater, T., Crane, J. and Pearce, N. 1998 Prevalence of adult asthma in relation to climate in New Zealand. *Environmental Health Perspectives,* **106**, 607–10.

Hastie, T. and Tibshirani, R. 1990 *Generalised Additive Models.* Chapman and Hall, London.

Ibrekk, H. and Morgan, M.G. 1987 Graphical communication of uncertain quantities to nontechnical people. *Risk Analysis,* **7**, 519–29.

Katsouyanni, K., Zmirou, D., Spix, C., Sunyer, J., Schouten, J.P., Pönkä, A., Anderson, H.R., Le Moullec, Y., Wojtyniak, B., Vigotti, M.A., Bacharova, L. and Schwartz, J. 1997 Short-term effects of air pollution on health: A European approach using epidemiologic time series data. The APHEA Project. Air Pollution Health Effects — A European Approach. *Public Health Reviews*, **25**, 7–18, discussion 19–28.

King, G. 1997 *A Solution to the Ecological Inference Problem; Reconstructing Individual Behavior from Aggregated Data.* Princeton University Press, Princeton.

Kuhn, L., Davidson, L.L. and Durkin, M.S. 1994 Use of Poisson regression and time series analysis for detecting changes over time in rates of child injury following a prevention program. *American Journal of Epidemiology,* **140**, 943–55.

Leigh, J., Macskill, P. and Nurminen, M. 1997 Revised quantitative risk assessment for silicosis and silica related lung cancer in Australia. *Annals of Occupational Hygiene*, **41**(Suppl. 1), 480–4.

Liang, K.Y. and Zeger, S.L. 1986 Longitudinal data analysis using generalised linear models. *Biometrika,* **73**, 13–22.

Maldonado, G. and Greenland, S. 1993 Interpreting model coefficients when the true model form in unknown. *Epidemiology,* **4**, 310–8.

McCullagh, P. and Nelder, J.A. 1983 *Generalised Linear Models.* Chapman and Hall, London.

McMichael, A.J. 1989 Setting environmental exposure standards: the role of the epidemiologist. *International Journal of Epidemiology,* **18**, 10–6.

Miettinen, O.S. 1985 *Theoretical Epidemiology. Principles of Occurrence Research in Medicine.* Delmar Publishers Inc., Albany, N.Y.

Morgenstern, H. 1982 Uses of ecologic analysis in epidemiologic research. *American Journal of Public Health,* **72**, 1336–44.

Morgenstern, H. and Bursic, E.S. 1982 A method for using epidemiologic data to estimate the potential impact of an intervention on the health status of a target population. *Journal of Community Health,* **7**, 292–309.

Morris, S. 1990 *Cancer Risk Assessment. A Quantitative Approach.* Marcel Decker, Inc., New York.

Navidi, W., Thomas, D., Stram, D. and Peters, J. 1994 Design and analysis of multilevel analytic studies with application to a study of air pollution. *Environmental Health Perspectives,* **102** (Suppl. 8), 25–32.

Nurminen, M. 1992 Assessment of excess risks in case-base studies. *Journal of Clinical Epidemiology,* **45**, 1081–92.

Nurminen, M. 1995a Linkage failures in ecological studies. *World Health Statistics Quarterly,* **48**, 78–84.

Nurminen, M. 1995b To use or not to use the odds ratio in epidemiologic analyses. *European Journal of Epidemiology,* **11**, 365–71.

Nurminen, M. 1997a On the epidemiologic notion of confounding and confounder identification. *Scandinavian Journal of Work Environment and Health,* **23**, 64–8.

Nurminen, M. 1997b Statistical significance a misconstrued notion in medical research. *Scandinavian Journal of Work Environment and Health,* **23**, 232–5.

Nurminen, M. and Nurminen, T. 1999 Methodologic issues in linking aggregated environment and health data. *Environmetrics,* **10** (in press).

Nurminen, M., Nurminen, T. and Corvalán, C.F. 1999 Methodologic issues in epidemiologic risk assessment. *Epidemiology,* **10**, 585–93.

Ostro, B. 1994 Estimating the health effects of air pollutants. A method with an application to Jakarta. Policy Research Working Paper 1301, The World Bank, Washington D.C.

Pearce, N. 1999 Epidemiology as a population science. *International Journal of Epidemiology,* **28**, S1015–8.

Plummer, M. and Clayton, D. 1996 Estimation of population exposure in ecologic studies. *Journal of the Royal Statistical Society B,* **58**, 113–26.

Poole, C. 1994 Ecologic analysis as outlook and method. (Editorial). *American Journal of Public Health,* **84**, 715–6.

Richardson, S., Stucker, I. and Hémon, D. 1987 Comparison of relative risks obtained in ecological and individual studies: some methodological considerations. *International Journal of Epidemiology,* **16**, 111–20.

Robinson, W.S. 1950 Ecological correlations and the behavior of individuals. *American Sociological Review,* **15**, 351–7.

Rothman, K.J. 1990 A sobering start for the cluster busters' conference. *American Journal of Epidemiology,* **132**(Suppl.), S6–13.

Rothman, K.J. 1993 Methodologic frontiers in environmental epidemiology. *Environmental Health Perspectives*, **101**(Suppl. 4), 19–21.

Selvin, H.C. 1958 Durkheim's "suicide" and problems of empirical research. *American Journal of Sociology,* **63**, 607–19.

Sheppard, L., Prentice, P.L. and Rossing, M.A. 1996 Design considerations for estimating exposure effects on disease risk, using aggregate data studies. *Statistics in Medicine,* **15**, 1849–58.

WHO (World Health Organization) 1995 Health and environment analysis and indicators for decision-making. *World Health Statistics Quarterly*, **48**, 70–170.

Zeger, S. and Qaqish, B. 1988 Markov regression models for time series: a quasi-likelihood approach. *Biometrics,* **44**, 1019–33.

Zidek, J.V., Wong, H., Le, N.D. and Burnett, R. 1996 Causality, measurement error and multicollinearity in epidemiology. *Environmetrics,* **7**, 441–51.

Chapter 7[*]

USING GEOGRAPHIC INFORMATION SYSTEMS TO LINK ENVIRONMENT AND HEALTH DATA

7.1   Geographical information systems

Analysing or interpreting the links between environment and health is, by its very nature, a spatial problem. Levels of risk vary geographically in response to variations in environmental conditions; health outcome and associated levels of need and health support vary as a consequence. Many of the questions facing the environmental epidemiologist and policy-maker are thus inherently geographical, and spatial analysis and mapping are vital components of their work. In research terms, these techniques provide an important step in the formulation and testing of hypotheses about links between environment and health. In policy terms, they are a valuable means of directing policy to areas and problems of greatest need, and of monitoring policy performance and effects.

Spatial analysis and mapping in environmental health have a long history. It is now traditional to trace their origin back at least as far as the seminal study of John Snow on cholera in London (Snow, 1855). Until recently, they could only be carried out manually, or using relatively simple mapping packages. Over the last ten years, however, the capability for spatial data manipulation has been revolutionised by the development of GIS. These systems have made mapping and many spatial analytical techniques much more readily available and enabled data visualisation in map form. They have also stimulated a wide range of new research initiatives into spatial operations and concepts which have greatly advanced understanding of how to analyse and interpret spatial phenomena.

Geographical information systems can simply be described as systems for the collection, storage, manipulation and display of spatially-referenced data (Maguire *et al.,* 1991). As such they perform a wide range of different functions. In addition to data capture and cleaning, data integration and data search and retrieval, these features include:

---

*   *This chapter was prepared by D. Briggs and K. Field*

- Visualisation of data in map or other form (graphs, tables, etc.) either on screen or as hard copy.
- Spatial data manipulation: i.e. the geographical manipulation and transformation of the data (e.g. buffering, topological overlay).
- Spatial data analysis: i.e. the analysis of spatial patterns and relationships within the data (e.g. spatial regression, spatial interpolation).

This wide-ranging capability means that GIS provide powerful tools for both environmental and health research. In the area of health, for example, they are now increasingly being used for the purpose of needs assessment, resource allocation and service planning (e.g. Bundred *et al.*, 1993; Sillince and Frost, 1993; Todd *et al.*, 1994; Jones and Bentham, 1995; Love and Lindquist, 1995; Lovett *et al.,* 1998) and for disease mapping (e.g. WHO 1997). Similarly, GIS are well established in the area of environmental analysis and management (e.g. Briggs, 1995; Dalbakova *et al.*, 1998). The use of GIS in the area of environmental health has, perhaps, been more limited. It is now expanding rapidly, however, as knowledge about the technology expands into environmental epidemiology (Stringer and Haslett, 1991; Gatrell and Naumann, 1992; Briggs and Elliott, 1995; Dunn *et al.,* 1995; Elliott and Briggs, 1998; Gatrell and Löytönen, 1998). This chapter outlines and illustrates some of the potential uses of GIS for environmental health management and policy, as part of HEADLAMP-type applications.

7.2   Visualisation

The ultimate purpose of most studies using GIS is to aid understanding of spatial patterns and relationships. Maps are usually prepared to enable information on distances, directions and areal extent to be retrieved, patterns revealed, and spatial relationships identified. Examples might include maps of disease rates (e.g. standard mortality rates), pollution levels (e.g. concentrations of air pollution), population or the costs of health service treatment. However, the paper map as a medium for storage and presentation of spatial data is no longer the final product. As the functionality of GIS has improved over the last decade, so has the ability for non-cartographers to create maps to illustrate their data at various stages. The creation of on-screen maps has enabled interactive interrogation of spatial databases in GIS projects, allowing visualisation of data in different ways and at different points in the process of spatial analysis. Furthermore, it has led to the development of alternative methods of displaying data, such as three-dimensional plots and animated maps.

Visualisation of data in a GIS environment can be applied in three different situations (Kraak and Ormeling, 1996). Firstly, data can be explored to discern spatial patterns, often as a method of examining raw or unknown data. Secondly, spatial analysis techniques can determine the relationship between two or more "coverages" of data (e.g. an overlay operation may combine several data sets to determine the spatial relationship between them). Thirdly, the results of analysis can be presented to communicate insight into the spatial patterns and processes discerned. There are thus three different uses of visualisation in GIS: exploration, analysis and presentation.

Visualisation of spatial data at each of these stages requires decisions to be made to select the most appropriate map type to illustrate the data. This may, in part, be determined by the format of the original data (i.e. whether it is point or area based) and the scale at which the map is to be produced. Maps produced within GIS are, characteristically, "thematic" and they are almost always "special purpose" (Dent, 1998). That is, they depict themes of information (attribute data) within geographical base maps (spatial data), often concentrating on the distribution of a single attribute or the relationship amongst several. Such maps show inherent location but are more concerned with focusing on the nature of the distribution. Furthermore, they may be qualitative or quantitative in design. A qualitative thematic map illustrates location and shows spatial distribution of mapped phenomena or nominal data (e.g. the location of air pollution monitoring sites in an urban area). On such a map, the user would be unable to determine quantity except that shown by relative areal extent. Quantitative thematic maps display spatial characteristics of empirical data illustrating variation on the phenomena from place to place (e.g. the depiction of levels of pollution measured at monitoring sites). In this sense, thematic maps have also been termed "statistical maps" because they invariably provide a summary measure of the theme of interest. The aim of an effective thematic map is to enable the user visually and intellectually to integrate both the spatial data and the thematic overlay.

Once the map type has been determined, effective use of "cartographic grammar" ensures that correct meaning is discernible from the data. Cartographic rules are available to assist in this process (Robinson *et al.*, 1995) but because they are not generally part of GIS software it is entirely possible that ineffective, or inaccurate, maps may be produced. Rules of generalisation (classification, simplification, exaggeration, symbolisation and induction) are designed to enhance map communication, with each mapped feature and its attributes contributing to the overall message and each appearing in its correct place in the visual hierarchy of design. The potential for ineffective mapping is most acute at the presentation stage where it is entirely possible for maps to be created which, visually, tell a different story from that intended. Care is always required in both designing and interpreting maps, because they are neither neutral nor passive instruments. Choice of colours, symbols, class intervals, typography, spatial units, map scale, projection and map content all influence the message that the map conveys, often subliminally (Smans and Estève, 1992; Monmonier, 1996). For example, red is often implicitly interpreted as meaning "bad", while green is interpreted as "good". Simply reversing a red–green colour scheme on a map can thus change its message totally. Whilst the creation of maps within a GIS is, in many ways, therefore a straightforward procedure, it is one which must be considered with due care. Most GIS afford a basic cartographic toolbox to aid data classification, symbolisation, legend depiction, etc. but these must be used with caution.

7.2.1   Consequences of map choice

As mentioned, the choice of map type to display thematic data can have a dramatic effect on the illustration of the spatial phenomena. Dot maps have long been used to map the statistical surface, with a single dot indicating location and, possibly, further classified to represent a set number of instances of the mapped feature. The nature of the distribution is indicated through accurate dot placement and the dot map can illustrate the character of a geographical distribution better than any other thematic map type (Dent, 1998). Variations in the spatial pattern, such as linearity and clustering, are discernible (although they need to be interpreted with great care) and the dot map also provides a suitable tool with which to gain an impression of relative density (albeit on an ordinal scale). Dot maps can be created within GIS in two ways. Firstly, where the data coverage is point-based, the location of the dots is implicit and dots can be measured based on one or more of the data attributes (which can also be used as a basis for creating proportional symbol maps). Secondly, the magnitude of a variable measured within areas can be illustrated through the dot technique. However, this second application must be treated with caution because dots are invariably randomly placed within the original areas. In this sense, the map implies location but the dots are only indicative of summary data for the area.

Because the simple dot map is essentially limited to illustrating location, other map types must be used to represent the often complex attributes associated with each point. Proportional symbol maps may be used, and they are conventionally available as part of GIS cartographic toolboxes, but there are a number of difficulties associated with the map type. These include problems of choosing appropriate symbol sizes and placement (e.g. to avoid symbol overlap) at the design stage, and perceptual difficulties in the interpretation process. For these reasons, point-based data are often depicted as a surface (e.g. isarithmic maps) or aggregated into areal units for illustrative purposes. In so doing, the data are conventionally depicted using a choropleth technique. The choropleth map is universally popular and is one of the fundamental map types available within GIS. Given its wide usage it would be easy to assume that it is also stable in its depiction of data. This is not necessarily the case. A map of health outcome using point data to represent individual cases might convey a very different picture, for example, to a choropleth map, in which the data have been aggregated into areal units (Figure 7.1). In essence, the choropleth map depicts an average count for each area because it has to be manipulated to take into account varying sizes of enumeration area. Furthermore, the choropleth map is not able to illustrate clustering, particularly in boundary areas, which may not have any association to the arbitrary enumeration area used for illustration.

7.2.2   Consequences of generalisation

There are many instances where it is prudent to generalise a data set to illustrate a spatial pattern effectively (e.g. aggregating census enumeration areas into larger entities). This is a common cartographic practice, particularly where the graphic limits of a map page prevent depiction of a large amount of detailed, dense spatial data. However, simplifying the spatial units of a data set within a GIS can have ramifications at the exploratory, analytical and presentational stages of visualisation.

A high degree of accuracy is implied through the use of GIS that can often mask generalised data. In terms of exploring or analysing data, the extent to which the framework for data collection and representation influences the meaning should be considered. Figure 7.2 shows two maps of the Townsend Deprivation Index. This index is a measure of the relative extent of material disadvantage, based on census data, which is often used in health studies to discern the relationship between pockets of deprivation and ill health (see also Chapter 3). The more detailed map is based on enumeration districts (EDs) (the finest level of spatial unit), with census data being collated at this level, classified and presented as a choropleth. For EDs, the maximum Townsend score is 8.6 (indicative of high levels of relative deprivation) and the minimum is –7.5. When the same census variables are collated at ward level (the next level of aggregation) the maximum ward score is 8.5 and the minimum is –4.0. Whilst there is little variation in the maximum scores between the maps there is a clear difference in the minimum score. The effect of aggregating data into larger spatial units effectively averages the data across all EDs for each ward, potentially masking important hotspots in the data set which could be important in exploratory and analytical terms — a problem referred to as the Modifiable Areal Unit Problem or MAUP (Openshaw, 1984).

7.2.3   Consequences of variation in unit-area size

Many data sets are derived (or presented) in areal form, as discussed above. This is useful, because such maps are easily recognisable and interpretable, but it does create difficulties due to the inherent spatiality of areal data. In particular, variation in unit-area size across a dataset has implications for the manipulation of data and its interpretation.

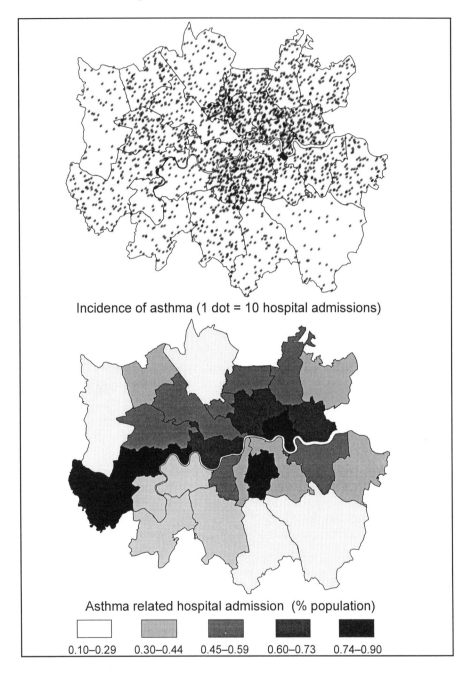

Incidence of asthma (1 dot = 10 hospital admissions)

Asthma related hospital admission (% population)

0.10–0.29	0.30–0.44	0.45–0.59	0.60–0.73	0.74–0.90

Figure 7.1 Comparison of dot and choropleth techniques illustrating incidence of asthma mapped for District Health Authorities of Greater London, UK (based on data from Field, 1998)

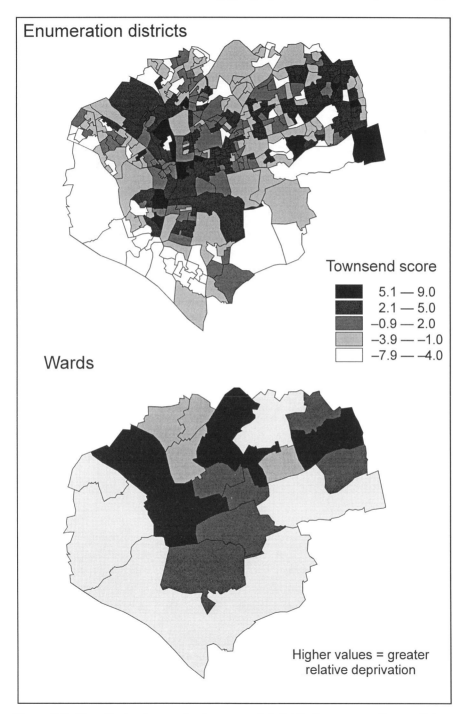

Figure 7.2 The effects of generalisation on spatial distribution: aggregating spatial units (based on data from Field, 1998)

Because unit-areas vary in size, it is not possible to map total values using a choropleth method. The difference in size of areas and their mapped values creates inaccurate impressions of the distribution; mapping data totals masks even densities or uniform distributions. It is therefore conventional to manipulate data and derive ratios involving area (e.g. persons per square km) or ratios independent of area (e.g. standard mortality ratio) in order to standardise the data and allow the effective visualisation of the density of data. Care is needed in these cases to ensure that the denominator used is appropriate. When using standard mortality ratios, for example, it may be more appropriate to stratify the data by age because mortality rates often vary markedly across different age groups.

In an interpretative sense, where unit-areas vary greatly in size the spatial pattern may be especially difficult to discern even when densities have been mapped. Larger areas tend to dominate the map and command attention, potentially drawing attention away from areas with more pressing concerns. This is a common problem in mapping disease rates across urban and rural areas. Generally, administrative units are considerably larger in the rural areas than in inner city areas. Rural areas, however, may have smaller populations and lower, or statistically less stable, disease rates. The inner city areas, where rates of illness are higher and where the data are most reliable, are often those that are least visible on the map (Figure 7.3). One way of avoiding these problems is to produce what are known as cartograms. These are maps in which the relative sizes of the constituent areas have been adjusted to help highlight the areas of greatest interest (Dorling, 1996). The spatial distortion inherent in such maps, however, has made them difficult to interpret and they have not been widely used.

7.3   Spatial data manipulation

The linkage of environment and health data relies heavily on the ability to manipulate spatial data. Many different methods of spatial data manipulation are available in GIS. They include methods such as buffering, topological overlay and spatial data integration. Buffering can be used, for example, to define and map areas of potential exposure around an emission source (e.g. a road, landfill site or chimney stack), based on distance from the source. Topological overlay techniques might then be used to intersect the resulting map with a map of population in order to assess the number of people potentially exposed. Alternatively, point-in-polygon techniques might be used to identify and count the number of cases of disease within the exposure zone.

One of the main uses of spatial data manipulation in environmental health studies is to convert the data to a common geographic framework. The need for this arises because many studies in environmental health inevitably involve the use of data based on different spatial units. Environmental data,

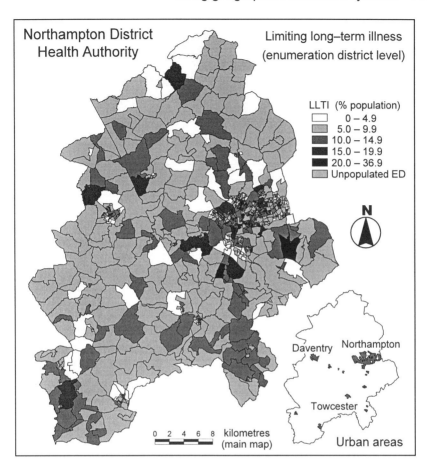

Figure 7.3 Rates of limiting long-term illness in Northamptonshire, UK mapped at ward level (based on data from Field, 1998)

for example, may relate to point measurements of pollution, made at individual sampling or monitoring points. Data relating to road traffic pollution or stream pollution may relate to line sources. Health data may comprise measures of disease or mortality rates aggregated to health service regions (e.g. health districts). Data on population or socio-economic conditions may be based on census regions. In order to analyse these different data sets together — for example to assess disease rates across the population or to relate levels of exposure to health outcome — the data need to undergo some form of spatial data integration.

Spatial data integration might take many different forms and be applied in many different ways, depending on the purpose of the analysis and the type of data involved. Figure 7.4 summarises the sorts of transformation that are

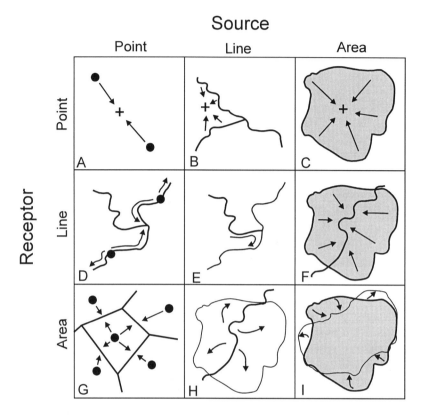

Figure 7.4 Examples of spatial data transformations in GIS. **A.** Point-to-point: prediction of levels of air pollution at a location between two monitoring stations. **B.** Line-to-point: prediction of levels of pollution at a location adjacent to a road network. **C.** Area-to-point: prediction of levels of pollution at a location within a mapped area of contamination. **D.** Point-to-line: estimation of the water quality of stream segments on the basis of data from monitoring sites. **E.** Line-to-line: extrapolation of water quality classes from classified segments to unclassified segments of the stream network. **F.** Area-to-line: prediction of levels of pollution along a stream or road passing through an area of mapped contamination. **G.** Point-to-area (tessellation): the mapping of pollution distribution on the basis of data from selected sample locations (e.g. soil sample sites). **H.** Line-to-area: the mapping of levels of pollution around a linear feature (e.g. roadway). **I.** Area-to-area: the interpolation of health data from one area base (e.g. health districts) to another (e.g. census districts) to allow matching against population data. Note that the arrows are not necessarily indicative of pollutant flow-lines, but they symbolise the process of spatial interpolation (After Briggs, 1992).

possible. In each case, integration involves the attempt to convert data relating to one set of spatial units (the source units) into another (the target units).

In some specific cases, the process of spatial transformation is little more than a process of summation. This might occur, for example, when the aim is to estimate disease rates on the basis of point data. In this case, point-in-polygon methods may be used to count the number of cases of the disease in each census district; the total can then be divided by the population of each census district to derive a rate. Another example is where data for small census districts can be aggregated to larger districts in which they are nested (e.g. to aggregate small area census-based population data to larger, and conformable administrative districts). Commonly, however, the process of spatial transformation is much more complex and relies on some degree of estimation or modelling of the spatial distribution of the data. Three common examples are:

- Where the need is to disaggregate data from larger source units into smaller target areas (e.g. to estimate small-area population characteristics from data available only at a more aggregated level).
- Where the aim is to convert one set of spatial units into another, non-conformable set (e.g. to match census districts to "natural" pollution zones).
- Where the spatial data are incomplete, so that it is necessary to estimate conditions at unsampled locations (e.g. to produce a map of pollution based on point measurements).

In each case, the accuracy of the transformation depends fundamentally on accuracy of the estimation method used. Table 7.1 lists some of the methods available. In the following sections these methods (and the issues involved) are discussed first in relation to area-based data (i.e. the conversion of data from one set of areal units to another, non-conformable set), and second in relation to point-based data (i.e. the construction of a continuous surface from sampled data).

7.3.1   The area-based problem

The spatial transformation of data from one set of area units to another is a common problem in environmental health assessments. It is illustrated here by considering the example of converting data on infant mortality from a relatively coarse set of health districts to a finer, and non-conformable set of census districts. The problem accords to that shown in "I" in Figure 7.4, and involves finding a way of modelling the spatial distribution of mortality within the source areas, so that the total number of cases for the new, target zones can be estimated.

Probably the simplest, and most often adopted, approach to this problem is to use some form of area-weighting. This is based on the assumption that the data are area-dependent (i.e. that the mortality rate is constant within each health district). Based on this assumption it is relatively simple to transform the data to the new spatial units. All that is required is to calculate the area of intersection between each source area and its overlapping target areas, and reallocate the mortality data on a proportional basis.

Table 7.1   GIS-based methods of spatial data integration

Operation	Technique	Description	Reference(s)
Spatial transformation of areal units	Areal weighting	Data from source zones are redistributed to target zones according to the proportionate area of overlap.	Deichmann, 1996; Fisher and Langford, 1995
	Modified areal weighting using control zones	Data from source zones are redistributed to target zones according to the proportionate area of the control zones selected.	Langford and Unwin, 1992; Moxley and Allanson, 1994
	Modified areal weighting using regression analysis	Data from source zones are redistributed to target zones using a weighting system derived from regression analysis.	Flowerdew and Green, 1992
Growth of new spatial units	Redistricting	New spatial units are generated by progressively combining adjacent units, according to predefined optimising criteria (e.g. to obtain areas of equal size).	Wise *et al.*, 1997
	Simulated annealing	As above, but allows backward steps in the process, thereby avoiding sub-optimal solutions.	Johnson *et al.*, 1989
Surface generation from area data	Bayesian map smoothing	Construction of "smoothed" maps from area data, taking account of uncertainty in the local area estimates.	Cressie, 1992; Mollie and Richardson, 1991
	Polygon filtering	A filter is passed iteratively over the map, and a weighted mean is calculated for each area from its preceding value and the values of all its neighbours.	Herzhog, 1989; Tobler, 1975

Continued

This assumption may be valid in some circumstances, but it is clearly liable to create significant errors either where the population itself is unevenly distributed within individual health districts, or where cases of infant mortality tend to be clustered. Better estimates might thus be made using alternative models of the spatial distribution of infant mortality. One simple improvement, for example, would be to assume that mortality is population, rather than area, dependent. In this case, mortality rates can be redistributed, not according to the proportionate area of overlap between the source and target zones, but between the proportionate population of these zones. This is relatively straightforward in the example presented here, because population data are likely to be available for the (target) census districts. If these data are stratified by age, then improved estimates might be possible; instead of the total population, data on the number of infants could be used.

Table 7.1   Continued

Operation	Technique	Description	Reference(s)
	Pycnophylactic interpolation	A lattice is laid over the map and grid cell values computed. Each grid cell is initially assigned a value based on the area in which it falls. A moving filter is then applied to produce a weighted average of its neighbouring values. After each iteration, the results are adjusted to ensure that the new values sum to the initial total.	Tobler, 1979
Surface generation from point data: local interpolation	Voronoi tessellation	Construction of irregular cells around each site, such that all points on the map are assigned the value of their nearest site.	Weibel and Heller, 1991
	TIN	Lines are created linking each point to its nearest neighbours in the form of a triangle. These form a continuous surface capable of further contouring or grid analysis.	Weibel and Heller, 1991
	Kriging	A suite of methods based on the construction of a semiovariogram, describing the relationship between the distance between each pair of data points and the difference in the measured values.	Isaaks and Srivastava, 1989; Oliver and Webster, 1990
Surface generation from point data: global interpolation	Trend surface analysis	A surface is fitted through the data points using regression techniques with locational (x and y) and attribute (z) values as inputs.	Haggett, 1968

Where population data are not directly available at an appropriate spatial scale, it may be possible to use other sources of data in order to simulate the population distribution at a finer spatial resolution, and to use this as a basis for reallocating the mortality. One especially useful approach in this context is to use land cover data (e.g. from aerial photographs or satellite data) as a proxy for population. At the simplest level of analysis, data on land cover may be interpreted using a binary classification to identify residential and non-residential areas (or built-up and open areas). Mortality data are then redistributed to the target zones according to the proportion of the residential areas that overlap with each source zone.

Where a more detailed classification of land cover is possible, this approach can be further refined by establishing differential weights (i.e. population densities) for each land cover class. Residential areas might be classified into high, medium and low-density housing, for example, each with a different assumed population density. These weights can then be used to redistribute the mortality data into the target zones. Where suitable population data exist, the weights can be derived not intuitively but by comparing the area of each land cover type with population size in each census district (e.g. using multiple regression techniques). If appropriate, other predictor variables could also be used in the analysis, such as level of poverty or housing conditions. Regression-based methods such as these need to be applied with care, however, because they can result in counterintuitive results: because of uncertainties in the regression models, for example, the mortality estimates for the target zones may no longer sum to the total for the region as a whole. Results thus need to be checked and adjusted to conserve the totals at all points, a process known as pycnophylactic interpolation (Tobler, 1979). Caution is also necessary because the assumptions used in modelling the spatial distribution may presume the effect of the very causal factors which are to be investigated. This would inevitably lead to marked bias in the estimates of health outcome and greatly distort subsequent analysis.

7.3.2 The point-based problem

The second case in which spatial transformation and interpolation are commonly necessary is in modelling spatial surfaces from sample (e.g. point) data. This is illustrated here by the attempt to estimate the air pollution surface across a city based on measured concentrations from monitoring sites. In this case, the need is to model the spatial pattern of pollutant concentrations at a fine resolution, so that reliable estimates of exposure can be made, either for point locations (e.g. places of residence) or for small areas (e.g. census districts). The problem is thus essentially one of spatial interpolation (i.e. the estimation of pollution levels at unsampled sites). Again, a range of methods are available within GIS for this purpose. These include the use of voronoi tessellation, triangulated irregular networks (TINs) and kriging.

Voronoi tessellation
One of the simplest and most widely available interpolation methods within GIS is voronoi tessellation. This involves the construction of polygons (voronoi) around each point (locus) such that every location within that polygon is closer to its own locus than to that of any other polygon (Figure 7.5). The measured value at each locus is then attributed to the polygon with which it is associated.

Voronoi tessellation is relatively simple, yet it is also conceptually crude. It relies wholly on proximity as the measure of affinity. It also assumes a disjunct pattern of variation in the variable of interest, such that conditions are uniform within any single polygon, but change abruptly at its border (Figure 7.6). Although this may have some degree of validity in the case of phenomena distributed on the basis of administrative areas (such as effects of General Practitioner (GP) diagnosis), it is rarely appropriate in the case of

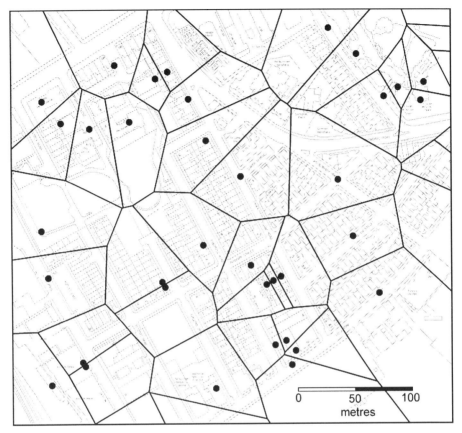

Figure 7.5 An example of voronoi polygons around air pollution monitoring sites in Kensington, London (courtesy of C. de Hoogh, University College Northampton)

environmental data, all of which tend to show much more complex and transitional patterns. Voronoi are also highly dependent on the distribution of the loci used to generate the polygons. Widely spaced loci have a larger effect on the outcome than densely spaced points. Odd-shaped polygons may also be formed at the margins of the study area, due to the absence of control points outside the area.

Triangulated irregular networks

Many GIS also provide the facility to generate surfaces using TINs. Triangulated irregular networks are created by constructing line segments, linking the point locations, to form triangles in which all sample points are connected with their two nearest neighbours. These triangles represent areas of uniform slope and can be of various (irregular) sizes. The gradient between each two points is assumed to be linear, although other models can be applied. The

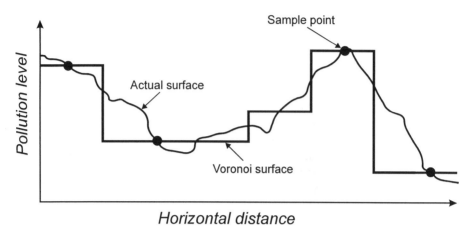

Figure 7.6   Assumed pattern of variation in voronoi tesselation.

triangulated network can be used to generate a number of derived surfaces (e.g. of gradient or direction of slope in the modelled variable). This surface may be represented either in raster form (i.e. as a grid of fine cells), by constructing contours or by creating a three-dimensional digital terrain model (DTM) of the continuous surface.

This method of contouring is, in fact, one of the most deeply established cartographic techniques for presenting continuous surfaces, such as pollution concentration fields. The technique is, nevertheless, not devoid of problems. The most serious limitation relates to the way in which the triangulation routine operates. With this method, contours are defined simply on the basis of the nearest three points. Points at greater distance are ignored, and there is consequently no weighting of the surface to allow for more general trends. The modelled surface is thus extremely sensitive to individual values, and it is not uncommon for large peaks or troughs to be formed around individual extreme data points. The shape of the surface is also influenced to a significant degree by the distance function chosen to represent the inter-site surface, and by the density and arrangement of the sampling points. As with voronoi tessellation, marked edge effects may also occur at the margins of the study area due to a lack of control points outside the convex hull of the triangulation network.

Kriging
The term kriging refers to a suite of related methods, all based on the principle that spatial variation in any phenomenon can be divided into three components: systematic trend (or drift), random but spatially-correlated variation (such that close points are more similar to each other than more distant points), and random spatially uncorrelated variation (noise). Oliver and Webster (1990) provide a valuable review of the method.

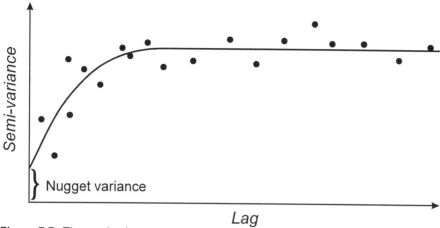

Figure 7.7 The semivariogram

The procedure involves calculating the semivariance: i.e. the relationship between inter-site distance (or lag) and the between site difference in the measured phenomenon. Plotting the semivariance typically produces a curve (the semivariogram) like that shown in Figure 7.7. The shape of the semivariogram indicates the nature of spatial variation involved. Often it is convex in form, the semivariance initially rising from some point above the origin, reaching a threshold (the sill) and then levelling off. The rising curve defines the spatially correlated variation in the data. The point of intersection on the y axis is known as the nugget variance and defines the "noise". A concave-upward semivariogram may indicate trend (drift) in the data. This normally needs to be removed prior to further analysis, for example using trend surface methods. The semivariogram is then recomputed using the detrended residuals.

Different models may be used to describe the semivariance quantitatively. Two groups of models are normally recognised: bounded models (characterised by a finite *a priori* variance, known as the sill) and unbounded models (lacking a distinct sill). Spherical and exponential models are the most common bounded forms used; power functions are the most widely used unbounded models. In either case, the choice of model may be extremely important, because it may significantly affect interpolation estimates. Best fit models are usually selected either by eye or by using least-squares methods.

Provided the nugget variance is not excessive, such that it dominates the spatial variation, the semivariogram can then be used to estimate values at unmeasured sites, based upon the contribution from surrounding points. The number of points contributing to the estimate can be determined by the user, taking account of the shape of the semiovariogram (i.e. the distance over which spatially correlated random variation effects can be detected). The

Table 7.2   Estimates of mean and 98th percentile NO_2 levels in Huddersfield using different interpolation methods

		Estimated concentrations ($\mu g\ m^{-3}$)	
Method	System	Mean	98th percentile
Arithmetic averaging	–	30.1	55.7
Voronoi tesselation	SPANS	27.6	56.2
	Arc/Info	30.2	46.2
TIN contouring	SPANS	27.9	46.1
Kriging	Arc/Info	29.1	40.1

Source: D. Briggs, Unpublished data

method can, however, be highly sensitive to the choice of model used to describe the semivariogram.

Although kriging is a relatively new technique, and one which has not yet been applied extensively in relation to environmental health assessments, it has a number of distinct advantages. First and foremost is the range of methods available, which allow interpolation to be adapted to the specific data and problem under consideration. In the form of punctual kriging it is what is known as an exact interpolator. By contrast, because it takes account of distance-related effects in the modelled surface, it is much less sensitive to individual data points than is the creation of a TIN. The procedure also generates an error estimate which allows the reliability of the modelled surface to be assessed across the mapped area.

Choice of interpolation method

These different methods of spatial interpolation clearly help to make GIS a powerful tool for mapping in environmental health. The range of methods available nevertheless imposes on the user an important question of choice. This is not a trivial decision, because the method of interpolation may have marked effects on the modelled outcome. Table 7.2, for example, shows estimates of the mean and 98th percentile concentration of NO_2 in Huddersfield, England. This is based upon a total of 79 sample points, within an area of about 300 km^2. The table compares four different methods of interpolation: arithmetic averaging (i.e. simple averaging of the data for the measured points), tessellation (area weighting) in both SPANS and Arc/Info packages, contouring (using a linear interpolation routine in SPANS) and kriging (using a circular semiovariogram in Arc/Info). As can be seen, considerable differences in the measured statistics occur, most notably in the 98th percentile value.

The performance of the different interpolation methods depends upon a number of factors including the nature of the underlying spatial variation in the phenomenon under consideration and the sample density and distribution. A number of comparative studies have been carried out (e.g. Dubrule, 1984; Laslett *et al.,* 1987; Abbass *et al.,* 1990) without clear consensus. In general, however, there are reasons to favour local methods of interpolation (such as kriging) over global methods (such as trend surface analysis) because the former are more sensitive to local variations in the data and thus do not produce as much smoothing of the modelled surface. Kriging also provides error estimates for the modelled surface.

7.4 Spatial data analysis

Maps are a valuable means of visualising spatial data on environment and health, but the information they contain is not always immediately apparent. Spatial data analysis is therefore an important step in exploring links between environment and health data. The aim of spatial data analysis is to test for patterns in the data, or for spatial relationships between different datasets.

7.4.1 Searching for point clusters

In the case of individual (i.e. point) data, spatial variation is likely to be expressed in a number of ways. Cases of interest may be more or less clustered, and they may be distributed systematically or randomly. Analysis of point data therefore commonly involves the search for clusters, and testing for spatial structure in the points (Kulldorff, 1998).

The search for clusters in point health data has proved to be especially controversial. It is an approach which has been motivated to a large extent by concerns about raised incidences of disease around industrial installations (e.g. nuclear power stations and processing plants, coking works). It is an approach, however, which presents a number of fundamental technical and statistical problems. These lie primarily in the statistical difficulty of testing for true excesses, and in the complexity of searching large areas, and large numbers of cases, for possible clusters.

Geographic information systems would nevertheless seem to offer a solution to the second of these problems, because they provide the facility for efficiently scanning data through moving window, buffering and point-in-polygon techniques. This approach was developed and applied by Openshaw *et al.* (1987) in establishing what was called a Geographical Analysis Machine (GAM). This enabled maps to be scanned systematically by constructing buffer zones around a fixed lattice of points in the study area. If the number of observed cases exceeded the expected then the circle was recorded. Following repeated scanning with circles of different radius, the results were mapped, and locations which provided the focus for a large number of overlapping circles were identified.

The Geographical Analysis Machine was originally developed as an exploratory and descriptive tool to help generate new hypotheses and direct research to specific areas. It was used, for example, to analyse excess cases of acute lymphoblastic leukaemia in children in Northern England (Openshaw *et al.*, 1987). The method has, however, attracted considerable criticism, not least because it involves double-counting of individual cases and the difficulty of analysing the resulting maps (e.g. Besag and Newell, 1991). Subsequent versions of GAM (Openshaw, 1990) resolved some of these problems, but have failed fully to satisfy its critics. In the meantime, various other methods have been suggested for analysing variations in health outcome. Besag and Newell (1991), for example, propose an alternative (and statistically more robust) method, in which the cumulative number of cases is counted with increasing distance from any point of interest, and the distribution compared with that of the population at risk.

7.4.2   Map smoothing
In the case of area data, similar issues occur. The basic question is to what extent the data show spatial structure, whether in the form of regional drift or trend or clustering. One recently developed approach to this question is provided by "map smoothing". This uses Bayesian statistical methods to determine whether there is any underlying structure (i.e. extra-Poisson variability) in the spatial distributions (e.g. Mollie and Richardson, 1991). It is based on the principle that spatial structure in the data will be evidenced by a tendency for nearby areas to be more similar to each other (in terms of health outcome) than more distant areas. Rates of disease are thus compared in terms of the adjacency or proximity of the map units.

To date, these techniques have not been fully integrated into GIS. Nevertheless, because they are based on the assessment of proximity or adjacency of the survey units, they can clearly benefit from the use of GIS. Thus Vincze *et al.* (1998) used map smoothing techniques in combination with GIS to investigate associations between iodide in drinking water and liver cirrhosis in Hungary. As part of the SAVIAH (Small Area Variation in Air quality and Health) study, the same methods were used to analyse patterns of asthma in urban areas, prior to examining links with air pollution (Elliott and Briggs, 1998). López-Abente (1998) also used Bayesian methods to examine relative risks of mortality due to connective tissue tumours, non-Hodgkin's lymphomas and multiple myeloma in Spain, at the province level. As commonly happens, use of Bayesian smoothing greatly reduced the range of relative risk (because the highest estimated risks in the raw data tended to be based on small numbers of cases, and were therefore unstable). For multiple myeloma, smoothing revealed a strong west-east gradient in relative risks, which was not immediately apparent in the raw data. A preliminary ecological analysis for the three health outcomes, using the smoothed rates, suggested associations with rates of pesticide application in agriculture — insecticides in the case of connective tissue tumours, molluscicides in the case of non-Hodgkin's lymphomas, and acaricides in the case of myeloma. Although no more than exploratory, the study thus indicated important lines for further research.

7.5   Conclusions

Geographical information systems clearly have much to offer in attempts to explore links between environment and health. They are powerful systems for the collection and integration of spatial and attribute data, and they provide a means for visualising and displaying these data in map form. They also offer a basis for statistical calculation and analysis of spatial data. As costs of purchase come down, and as more digital data sets become available, GIS are becoming more accessible to many users in environmental health. They are therefore increasingly becoming an integral part of environmental health analysis, and important tools for risk assessment, risk communication and risk management.

The very power and persuasiveness of GIS nevertheless mean that they need to be used with care. Their use in environmental health can do much to influence people's perceptions and actions. It thus behoves all who use GIS in these circumstances to pay particular attention to the quality of the data used. The old adage of "garbage in garbage out" is as true in GIS as in any other form of data analysis, but often less apparent because of the sophistication of the output, and the hidden complexity of the analytical operations involved. Many data used for GIS applications are nevertheless subject to significant errors and inconsistencies. Linking these to provide spatial coverage, or over-laying them to derive new information, may generate complex and unseen error surfaces. Data quality control is thus of the utmost importance. It must consider not only the spatial properties of the data (e.g. are points or lines located correctly) but also, and often more importantly, the quality of the attached attributes (e.g. are the values being mapped accurate and unbiased).

Sadly, data quality control is often inhibited by the poor documentation attached to many data sets. In many cases, it is difficult to obtain independent reference data against which to judge the quality of the data used or the results obtained. Manual data checking can also be extremely onerous. GIS methods themselves, however, are often of assistance in data checking. Mapping of each variable separately can provide a useful means of identifying hiatuses or outliers in the surfaces. Calculation of spatial statistics (e.g. point densities, averages for moving windows across the surface) can similarly indicate errors or discrepancies. Point in polygon searches can be a useful means of checking locational accuracy of data points (e.g. addresses, post codes).

In this context, also, issues of scale and resolution are of special importance. Data from GIS are often described as being "scale-free". In principle, this is true. Through the generalisation features available in GIS, for example, large scale (i.e. detailed) maps can be reduced to a smaller, and less detailed, scale. Conversely, the zoom facilities available in GIS allow maps to be easily enlarged for display either on screen or in hard copy. Generalisation, however, involves loss of detail in the data. Normally the data must be reduced in volume (i.e. weeded) during generalisation to remove redundant data and to prevent crowding and clogging of the output. Magnification of the data, in contrast, can increase the size at which maps are displayed, but cannot add new data points to the original set. It is thus impossible to improve upon the resolution of the data beyond that already stored in the database. The ultimate limits to map resolution are therefore determined by the scale and resolution of the original, source data.

Despite their visual impact, therefore, GIS cannot tell the whole story, and they are at the mercy of the data and the models which they use. Used appropriately, and with reliable data, they can undoubtedly aid and enhance attempts to understand relationships between environment and health and help to guide action to safeguard human health. But used incautiously, they can mislead.

7.6   References

Abbass, T. El, Jalloulli, C., Albouy, Y. and Diament, M. 1990 A comparison of surface fitting algorithms for geophysical data. *Terra Nova,* **2**, 467–75.

Besag, J. and Newell, J. 1991 The detection of clusters in rare diseases. *Journal of the Royal Statistical Society* A, **154**, 143–55.

Briggs, D.J. 1992 Mapping environmental exposures. In: P. Elliott, J. Cuzick and R. Stern [Eds] *Geographical and Environmental Epidemiology.* Oxford University Press, Oxford, 158–76.

Briggs, D.J. 1995 Building a geographical information system in the European Community: the CORINE experience. In: M.J.C. de Lepper, H.K. Scholten and R.M. Stern [Eds] *The Added Value of Geographical Information Systems in Public and Environmental Health.* Kluwer Academic Publishers, Dordrecht, 299–314.

Briggs, D.J. and Elliott, P. 1995 GIS methods for the analysis of relationships between environment and health. *World Health Statistics Quarterly,* **48**, 85–94.

Bundred, P., Hirschfield, A. and Marsden, J. 1993 GIS in the planning of health services in a District Health Authority. *Proceedings of the AGI Conference 1993,* Taylor and Francis, London,1.1.1–1.1.4.

Cressie, N. 1992 Smoothing regional maps using Empirical Bayes Predictors. *Geographical Analysis,* **24**, 75–95.

Dalbakova, D.L., Dimitrova, R.S., Boeva, B.P., Henriques, W.D. and Briggs, D.J. 1998 Tools for risk assessment: Geographic Information Systems. In: D.J. Briggs, R. Stern and T.L. Tinker [Eds] *Environmental Health for All. Risk Assessment and Risk Communication for National Environmental Health Action Plans.* NATO Science Series 2, Environmental Security - Vol. 49, Kluwer, Dordrecht, 133–46.

Deichmann, U. 1996 *Smart interpolation. A review of spatial interpolation, database design and modelling. Use of GIS in Agricultural Research.* Global Resource Information Database, United Nations Environment Programme, Nairobi.

Dent, B. 1998 *Cartography: Thematic Map Design.* Fifth Edition, McGraw-Hill, New York.

Dorling, D. 1996 *Area Cartograms: their Use and Creation.* CATMOG No. 59. Geo Books, Norwich.

Dubrule, O. 1984 Comparing splines and kriging. *Computers and Geosciences,* **10**, 327–38.

Dunn, C.E., Woodhouse, J., Bhopal, R.S. and Acquilla, S.D. 1995 Asthma and factory emissions in northern England: addressing public concern by combining geographical and epidemiological methods. *Journal of Epidemiology and Community Health*, **49**, 395–400.

Elliott, P. and Briggs, D.J. 1998 Recent developments in the geographical analysis of small area health and environmental data. In: G. Scally [Ed.] *Progress in Public Health.* FT Healthcare, London, 101–25.

Field, K. S. 1998 Modelling Health Care Utilization: an Applied GIS Approach. Unpublished PhD, Leicester University.

Fisher, P. and Langford, M. 1995 Modelling the errors in areal interpolation between zonal systems by Monte Carlo simulation. *Environment and Planning A*, **27**, 211–24.

Flowerdew, R. and Green, M. 1992 Developments in areal interpolation methods in GIS. *Annals of Regional Science,* **26**, 79–95.

Gatrell, A. and Naumann, I. 1992 Hospital location planning: a pilot GIS study. Unpublished report, North West Regional Research Laboratory, Lancaster University, Lancaster.

Gatrell, A. and Löytönen, M. 1998 *GIS and Health.* GISData 6. Taylor and Francis, London.

Haggett, P. 1968 Trend surface mapping in the comparison of inter-regional structures. *Regional Science Association*, **20**, 19–28.

Herzhog, A. 1989 Modelling reliability on statistical surfaces by polygon filtering. In: M. Goodchild and S. Gopal [Eds] *Accuracy of Spatial Databases.* Taylor and Francis, London, 209–18.

Isaaks, E.H. and Srivastava, R.M. 1989 *An Introduction to Applied Statistics*. Oxford University Press, Oxford.

Johnson, D.S., Aragon, C.R., McGeoch, L.A. and Schevon, C. 1989 Optimisation by simulated annealing: an experimental evaluation. Part 1. Graph partitioning. *Operations Research,* **37**, 865–92.

Jones, A.P. and Bentham, G. 1995 Emergency medical service accessibility and outcome from road traffic accidents. *Public Health,* **109**, 169–77.

Kraak, M.J. and Ormeling, F.J. 1996 *Cartography: Visualization of Spatial Data*. Addison Wesley Longman, Harlow.

Kulldorff, M. 1998 Statistical methods for spatial epidemiology: tests for randomness. In: A.A. Gatrell, and M. Löytönen [Eds] *GIS and Health*. GISData 6. Taylor and Francis, London, 49–62.

Langford, M. and Unwin, D.J. 1992 Generating and mapping population density surfaces within a GIS. *The Cartographic Journal,* **31**, 21–6.

Laslett, G.M., McBratney, A.B., Phalli, P.J. and Hutchinson, M.F. 1987 Comparison of several spatial prediction methods for soil pH. *Journal of Soil Science,* **38**, 325–70.

López-Abante, G. 1998 Bayesian analysis of emerging neoplasms in Spain. In: A. A. Gatrell and M. Löytönen [Eds] *GIS and Health*. GISData 6. Taylor and Francis, London, 125–37.

Love, D. and Lindquist, P. 1995 Geographic accessibility of hospitals to the aged. *Health Services Research,* **29**(6), 629–51.

Lovett, A., Haynes, R., Bentham, G., Gale, S., Brainard, J. and Suennenberg, G. 1998 Improving health needs assessment using patient register information in a GIS. In: A.A. Gatrell and M. Löytönen [Eds] *GIS and Health*. GISData 6. Taylor and Francis, London, 191–203.

Maguire, D.J., Goodchild, M.F. and Rhind, D.W. [Eds] 1991 *Geographical Information Systems. Volume 1. Principles*. Longman Scientific and Technical, London.

Mollie, A. and Richardson, S. 1991 Empirical Bayes estimates of cancer mortality rates using spatial models. *Statistics in Medicine,* **10**, 95–112.

Monmonier, M. 1996 *How to Lie with Maps*. University of Chicago Press, Chicago.

Moxley, A. and Allanson, P. 1994 Areal interpolation of spatially extensive variables: a comparison of alternatives. *International Journal of Geographical Information Systems,* **8**, 479–87.

Oliver, M.A. and Webster, R. 1990 Kriging: a method of interpolation for geographical information systems. *International Journal of Geographical Information Systems,* **4**, 313–32.

Openshaw S. 1984 *The Modifiable Areal Unit Problem*. CATMOG No 38, Geo Books, Norwich.

Openshaw, S. 1990 Automating the search for cancer clusters: a review of problems, progress and opportunities. In. R. Thomas [Ed] *Spatial Epidemiology.* Pion, London, 48–79.

Openshaw, S., Charlton, M., Wymer, C. and Craft, A.W. 1987 A mark 1 Geographical Analysis Machine for the automated analysis of point data sets. *International Journal of Geographical Information Systems,* **1**, 335–58.

Robinson, A.H., Morrison, J.L., Muehrcke, P.C., Kimerling, A.J. and Guptill, S.C. 1995 *Elements of Cartography.* Sixth Edition, John Wiley & Sons, New York.

Sillince, J.A.A. and Frost, C.E.B. 1993 Management information systems in UK primary health care: the need for a strategy. *International Journal of Information Management*, **13**, 425–37.

Smans, M. and Estève, J. 1992 Practical approaches to disease mapping. In: P. Elliott, J. Cuzick and R. Stern [Eds] *Geographical and Environmental Epidemiology*. Oxford University Press, Oxford, 141–50.

Snow J. 1855 *On the Mode of Communication of Cholera.* Second edition, Churchill Livingstone, London.

Stringer, P. and Haslett, J. 1991 Exploratory, interactive analysis of spatial data: an illustration in the area of health inequalities. Unpublished report. Northern Ireland Regional Research Laboratory, Trinity College, Dublin.

Tobler, W.R. 1975 Linear operations applied to areal data. In: J.C. Davis and J.A. McCullagh [Eds] *Display and Analysis of Spatial Data.* Wiley, London, 14–37.

Tobler, W.R. 1979 Smooth pycnophylatic interpolation for geographical regions. *Journal of the American Statistical Association*, **74,** 519–30.

Todd, P., Bundred, P. and Brown, P. 1994 The demography of demand for oncology services: A health care planning GIS application. In: *Proceedings of the AGI Conference 1994*, Taylor and Francis, London, 17.1.1–17.1.7.

Vincze, I., Elek, G. and Nador, G. 1998 Is iodide a confounding or effect modifying factor of liver cirrhosis? In: D.J. Briggs, R. Stern and T.L. Tinker [Eds] *Environmental Health for All. Risk Assessment and Risk Communication for National Environmental Health Action Plans.* NATO Science Series 2. Environmental Security - Vol. 49. Kluwer, Dordrecht, 77–84.

Weibel, R. and Heller, M. 1991 Digital terrain modelling. In: D.J. Maguire, M.F. Goodchild and D.W. Rhind [Eds] *Geographical Information Systems. Volume 1. Principles.* Longman Scientific and Technical, London, 269–97.

WHO 1997 *Atlas of Mortality in Europe. Subnational Patterns 1980/1981 and 1990/1991.* WHO Regional Publications, European Series No. 75, World Health Organization, Geneva.

Wise, S., Haining, R. and Ma, J. 1997 Regionalisation tools for the exploratory spatial analysis of health data. In: A. Getis and M.M. Fisher [Eds] *Recent Developments in Spatial Analysis*. Springer Verlag, Berlin, 83–100.

Chapter 8[*]

APPLICATION OF HEADLAMP IN THE FIELD

8.1 Introduction

In 1995, six HEADLAMP field studies were conducted in selected developing country cities: Calcutta (India), Cape Town (South Africa), Cotonou (Benin), Managua (Nicaragua), Manila (the Philippines) and Talcahuano (Chile). These cities were chosen as a wide representation from developing countries in different world regions. Of most interest were cities in countries experiencing a combination of traditional and modern environmental health risks. As societies develop, modern hazards such as air and chemical water contamination and solid hazardous waste accumulation tend to increase, while traditional hazards such as lack of safe drinking water and basic sanitation and indoor air pollution from the use of coal and biomass fuel tend to decrease. This has been termed the environmental health risk transition (Kjellström and Rosenstock, 1990; Smith, 1990, 1997). The final selection of study sites was dictated by the existence and knowledge of a reputable research centre in each city and the identification of an investigator who would take responsibility for the local co-ordination of the study. Because the range chosen is wide, the characteristics of these places vary considerably (see Figure 8.1). Differences between the cities are further illustrated by the Human Development Index (UNDP, 1990) for each parent country (Table 8.1). This index includes a health related indicator (life expectancy at birth), and two important health determinants, namely educational attainment (measured as adult literacy and school enrolment ratios) and standard of living (measured as real GDP per capita).

In each city, the field studies were designed to meet the following objectives:

- To identify the specific local environmental health problems that pose a threat to human health.
- To describe the local decision-making process in environmental health.
- To test the application of the proposed HEADLAMP methods.
- To field test a proposed set of environmental health indicators.

---

[*]   *This chapter was prepared by C. Corvalán, G. Zielhuis and F. Barten*

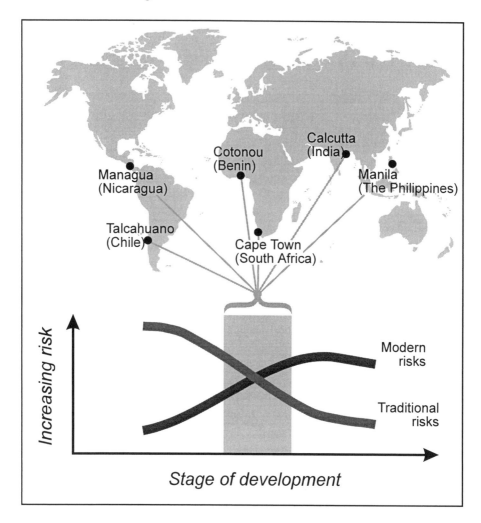

Figure 8.1   HEADLAMP field study sites and their place in the environmental health risk transition (After Smith 1990, 1997; WHO, 1997)

As a background to the studies, a draft version of the HEADLAMP report (Briggs *et al.*, 1996; Corvalán *et al.,* 1997), outlining the methods which might be used, was provided to the study researchers. Based upon this report, a set of field study guidelines was established for each area (see section 8.3). These were designed to be flexible and realistic in relation to the available financial resources and the anticipated availability of data. Given the geographical and developmental differences in the sites chosen, it was expected that the findings would be generally applicable, with caution, to other cities of similar characteristics, at least within the regions they represented.

Table 8.1   Human Development Index and its health-related components for parent countries, 1994

Country	Human Development Index	Life expectancy at birth (years)	Adult literacy rate (%)	School enrolment rate[1]	Real GDP per capita (PPP$)
Benin	0.368	54.2	35.5	35	1,696
India	0.446	61.3	51.2	56	1,348
Nicaragua	0.530	67.3	65.3	62	1,580
The Philippines	0.672	67.0	94.4	78	2,681
South Africa	0.716	63.7	81.4	81	4,291
Chile	0.891	75.1	95.0	72	9,129
Developing countries	0.576	61.8	69.7	56	2,904
Developed countries	0.911	74.1	98.5	83	15,986

[1]   First, second and third levels combined   Source: UNDP, 1997

8.2   From the global to the local perspective

At the national and, to a large extent, the global level, a considerable volume of information tends to be available on specific health issues and the impact of the environment on health. At the local level, in contrast, such information is often scarce or difficult to obtain. The overall interpretation is derived not by aggregation of, or extrapolation from, local knowledge but by collection of statistics at the broad regional or national scale. Even when derived from more detailed, sub-national data, the reported national indicators are usually presented as averages, which thus disguise important socio-economic, ethnic, geographical or gender differences. More usually, disaggregated, local data (e.g. at the level of individual cities) are not available at all. As a result, important information on patterns of health status and on its relationship with environmental, social and policy factors tend to be hidden. Two examples illustrate the case:

- *Example 1. Infant mortality rates in Chile.* Graph A in Figure 8.2 shows the national reported rates of infant mortality for selected years, between 1970 and 1995. Graph B shows rates for the same years for each of the 12 provinces and for the metropolitan region (Santiago). The national indicator clearly shows that infant mortality rate has dramatically decreased in recent decades. The regional indicators reveal more. While infant mortality has declined in all areas, it has done so most markedly in the areas with the highest rates. Thus the variation in rates between regions has been reduced. This latter trend can be interpreted as a result of the reduction in social inequalities between the regions. Undoubtedly, these data do not tell

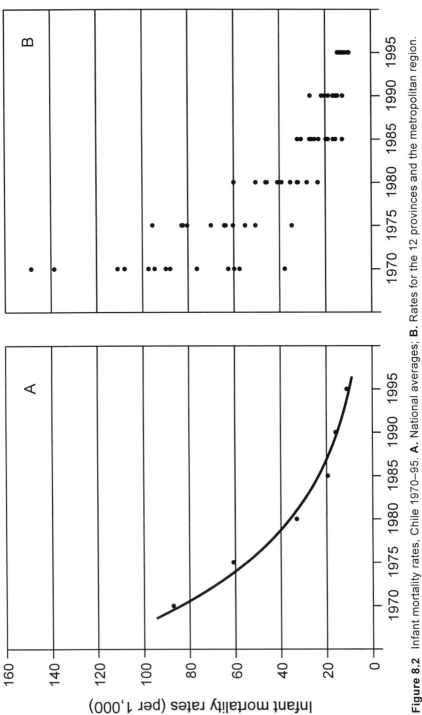

Figure 8.2 Infant mortality rates, Chile 1970–95. **A.** National averages; **B.** Rates for the 12 provinces and the metropolitan region. (After Instituto Nacional de Estadísticas 1977, 1985, 1991, 1992, 1995)

the whole story: even the regional data are still geographic aggregations which hide rural–urban, intra-urban and probably other differences. Nevertheless, the regional data do imply that even these differences may not be as wide as they were in the past. By examining and comparing the regional data, therefore, it is possible to begin to form hypotheses about the trends and processes in operation, and to decide where to look in more detail to investigate or detect specific situations.

- *Example 2. Environment and disability free survival.* In a wide-ranging study, Murray and Lopez (1994, 1996) proposed a methodology for estimating the total "burden of disease", by combining information on the number of deaths with the impact of disability in a population. The resulting measure has been referred to as Disability Adjusted Life Years (DALYs). Initial estimates were first published in the 1993 World Development Report, *Investing in Health* (World Bank, 1993). As part of this work, several "risk factors" were considered, and calculations were made of their contribution to the global burden of disease (Murray and Lopez, 1996). More recently, WHO presented estimates of the approximate proportion of DALYs associated with environmental factors, by assigning weight factors in the form of the estimated environmental fraction to reported DALYs for relevant diseases (Table 8.2) (WHO, 1997).

According to these estimates 23 per cent of the global burden of disease is related to environmental factors. Murray and Lopez (1996) present data on the numbers of DALYs for major disease categories by eight world regions, four age groups and gender. Using this, it is possible to make more detailed regional estimates of the impact of the environment on health, as shown in Table 8.3. This table reveals two important observations. First, the regional differences in DALYs are large — up to 80 times higher in Sub-Saharan Africa than in the developed market economies. Secondly, most of the environmental burden of disease in the less developed regions falls on children under five years of age (for a more detailed analysis, see Smith *et al.*, 1999). This example thus highlights some of the uncertainties that may arise from using aggregated (national or global) data and indicates some of the differences that may be encountered when comparing cities or regions in developing countries.

8.3   Field studies guidelines

This section provides an abridged version of the field study guidelines (WHO, 1995) used to prepare the studies described in section 8.4. The guidelines are divided into four parts (corresponding to the four main objectives of the study), each presented with details of the procedures involved and their expected outputs (Box 8.1). A list of suggested health-related indicators was also proposed (Box 8.2).

Table 8.2 Proportion of global DALYs associated with environmental exposures, 1990

Disease	Exposure situation A	E	W	F	H	DALYs (10³)	Environmental fraction (%)	Environmental DALYs (10³)	Percentage of all DALYs
Acute respiratory infection	✓				✓	116,696	60	70,017	5.0
Diarrhoeal disease		✓	✓	✓		99,633	90	89,670	6.5
Vaccine-preventable infections		✓	✓	✓		71,173	10	7,117	0.5
Tuberculosis					✓	38,426	10	3,843	0.3
Malaria			✓		✓	31,706	90	28,535	2.1
Injuries, unintentional					✓	152,188	30	45,656	3.3
Mental health					✓	144,950	10	14,495	1.1
Cardiovascular disease	✓					133,236	10	13,324	1.0
Cancer	✓			✓		70,513	25	17,628	1.3
Chronic respiratory disease	✓					60,370	50	30,185	2.2
Total, above diseases	–	–	–	–	–	918,891	35	320,470	23.0
Other diseases	–	–	–	–	–	460,347	–	–	–
Total, all diseases	–	–	–	–	–	1,379,238	23	320,470	–

Source: Adapted from WHO, 1997

A   Polluted air
E   Excreta/wastes
W   Water
F   Food
H   Unhealthy housing

Table 8.3   Environmental DALYs by world region

World region	Environmental DALYs as a percent of all DALYs		Environmental DALYs per 10^3 persons	
	Age 0–4	Age 5+	Age 0–4	Age 5+
Developed market economies	0.5	13.7	9	18
Former USSR and Eastern Europe	1.3	13.5	30	26
China	5.7	13.7	100	28
Latin America and the Caribbean	8.0	9.9	138	25
Other Asia and Islands	12.3	10.3	254	31
Middle East and North Africa	16.0	6.9	297	25
India	15.3	9.9	376	39
Sub-Saharan Africa	22.5	8.6	705	62

Source: Calculated from data in WHO, 1997 and Murray and Lopez, 1996

The process to develop an environmental health profile in each study area consisted of a combination of:

- A literature review of both published and unpublished reports at different levels (local, regional and national); and
- The identification of all existing health, environment and demographic data and their sources.

Where possible, this was complemented with interviews with key persons from government, community and NGOs. In some sites, workshops were conducted as part of the field study and these brought together representatives from the different sectors with an interest in any relevant aspect of health and the environment.

An extensive report was prepared for each field study. An interim evaluation of the studies was made at a meeting with all field study research co-ordinators in late 1995. Section 8.4 provides summary descriptions and findings from the six sites, together with an evaluation, based on these criteria. Section 8.5 presents an evaluation of the field studies based on the final reports.

8.4   Field applications

8.4.1   Introduction

Initial field studies were carried out in 1994 in Accra, Ghana (Songsore and Goldstein, 1995) and Sao Paulo, Brazil (Stephens *et al.*, 1995). These had the purpose of examining data availability and quality, and investigating the

Box 8.1   Guidelines for field studies

Objective
A. Identification of specific local environmental health problems which pose a threat to human health.
Procedure
- Describe the geographical area and population covered. This should include relevant demographic statistics (e.g. population size by age group and gender; population change over time, etc.).
- Describe the general health status of the population (e.g. infant mortality rate by gender; life expectancy at birth by gender, maternal mortality, etc.).
- Describe local environmental health problem areas: water, food, soil, shelter, air, workplace and transboundary or global problems affecting the local area under study.
- Describe general environmental health issues, actions and capabilities at the local level.
- In identifying each relevant indicator, discuss their availability, access and sources of health and environmental monitoring data (who owns these data, who has access to them, their quality and completeness). Identify current local use of indicators. If data are unavailable or insufficient, identify the needs and feasibility of collecting extra data. Include comments on identification of data format (e.g. hand written, typed or electronic records) and, if relevant, their potential transferability to electronic media. Where relevant, obtain raw data (from data providers) for analysis as part of this (or a future) study.
- Identify and contact government and non-government agencies that provide services (or that have policy/planning responsibilities) related to each environmental problem. Obtain from them relevant information and provide them with information on this project. Where relevant and possible, a workshop bringing all stakeholders together should be conducted.
Output
The output of this step is an environmental health profile of the study area.

Objective
B. The information and decision-making process using indicators.
Procedure
- Describe the local decision-making process for health and environment related problems.
- Describe the use of local indicators and of data linkage in the decision-making process.
- Identify relevant aspects of information and decision-making that lead to action. If action is lacking, discuss the major impediments for both decision-making and for implementation of decisions (i.e. action).
- Discuss existing intersectoral collaboration and/or impediments to such collaboration.
- Identify government agencies, institutions and organisations that have interest in the development of indicators and the ability to implement HEADLAMP methods.
Output
The output of this step is a description of the local information and decision-making process in relation to environmental health problems.

Continued

Box 8.1   Continued

Objective
C. Test the application of the methods proposed by HEADLAMP.
Procedure
- Describe the feasibility of data linkage and identification of the most appropriate methods for each environmental health problem identified in part A, above. Identify current use of data linkage, use of indicators and the decision-making process in each case. At this stage it would be useful to outline other potential methods to estimate the health impacts and discuss potential difficulties anticipated. Discuss any other major potential problems and provide suggestions for overcoming them.
- Identify the best forms of data presentation to aid the decision-making process in this particular setting. Identify other conditions that may be required for action in this situation. Seek feedback from a wide range of persons (government and NGOs) regarding the potential application of HEADLAMP for the needs and priorities of each area. Identify local needs, views and concerns. Identify local institutions and researchers working on epidemiological studies of the health effects of environmental contamination, on data linkage, and on developing indicators. If no local applications exist, discuss the feasibility of applying HEADLAMP methods in the form of ecological linkage, risk analysis or descriptive assessment using GIS.

Output
The output of this step is a discussion on the use of HEADLAMP methods in the study area and the identification of field based examples of data linkage.

Objective
D. Field test of a proposed set of Environmental Health Indicators (EHIs) for decision-making
Procedure
- A list of selected environmental health indicators and sustainable development indicators related to health (Box 8.2) needs to be tested at the local level. These indicators form part of a proposed list of sustainable development indicators related to Chapter 6 of Agenda 21, "Protecting and promoting human health" (United Nations, 1993) (for detailed methodology sheets for many of these and other indicators see United Nations, 1996, and Annex 1 of this volume).
- From the indicator list given in Box 8.2, indicate: which indicators are in use; which are not in use but would be easy to collect; and which would require a major effort to collect.
- If possible and relevant, give comments about each indicator in Box 8.2 (e.g. issues of quality, access, completeness, data format, and actual value).
- In part A, a series of environmental health issues in the study area are identified. Use Table 8.4, to the extent possible, to identify all relevant indicators related to each environmental health issue. Table 8.4 follows the Driving force, Pressure, State, Exposure, Effect, Action (DPSEEA) framework to classify existing (or feasible to collect) environmental health indicators.

Output
The output of this step is an assessment of the feasibility of collecting and using basic indicators and the identification of additional indicators that are specific to the study area.

Table 8.4   Template for environmental health indicators identified during the field study and classified using the DPSEEA framework

Environmental health issue:

Stage	Process	Descriptive indicator	Action indicator
Driving force	Type of development or human activities		
Pressure	Source activity		
	Emissions		
State	Environmental levels		
Exposure	Human exposure		
	Dose		
Effects	Early/sub-clinical		
	Moderate/clinical		
	Advanced/permanent		

The term "descriptive indicator" refers to indicators which describe the driving forces, pressures, state, exposures and effects; "action indicators" relate specifically to the actions linked to each of these steps (see also Table 3.1)

feasibility of linking health and environment data in each study area. In 1995 these pilot studies were followed by six field studies in Calcutta, India (Mukherjee *et al.*, 1995), Cape Town, South Africa (Lewin, 1995, 1996), Cotonou, Benin (Soton *et al.*, 1995, 1997), Managua, Nicaragua (Gonzalez *et al.*, 1995), Manila, the Philippines (Torres and Subida, 1996), and Talca-huano, Chile (Salinas *et al.*, 1995). For the purpose of this evaluation, only the 1995 studies are considered, because these share the same study protocol.

This evaluation is an attempt to assess the extent to which the HEADLAMP methodology worked in the field in these study areas. It should be kept in mind that none of these studies attempted to set in motion a full HEADLAMP project in their local settings. The main purpose of the studies was to see if the necessary preconditions existed for HEADLAMP activities to be developed. As such, they were restricted to research activities, without involving the commitment from policy makers at this early stage (Corvalán and Kjellström, 1997).

The following sections give brief descriptions of the field studies in each of the six cities used for this evaluation. These descriptions are based on final reports obtained in 1995–97. An attempt has been made to maintain comparability between cities, but the summaries presented here are based only on the

---

**Box 8.2   List of selected environmental health indicators and
                    sustainable development indicators related to health**

- Life expectancy at birth
- Infant mortality rate (per 1,000 live births)
- Maternal mortality rate (per 1,000 live births)
- Percentage of the population with access to sufficient quantity of safe drinking water
- Percentage of the population with access to hygienic excreta disposal facilities
- Percentage of people served by public waste removal service
- Percentage of people exposed to high concentrations of health damaging air pollution (outdoors and indoors separately)
- Percentage of people without access to adequate food supply
- Percentage of new-born weighing at least 2,500 g at birth
- Percentage population covered by primary health care
- Percentage of the eligible population that have been fully immunised according to national immunisation policies
- Prevalence of malaria
- Incidence of acute enteric infections
- Prevalence of intestinal helminths among children (aged 2 to 15)
- Adult literacy rate
- Percentage Gross National Product spent on health
- Percentage of national health expenditure devoted to primary health care, health centres and regional hospitals

---

information collected during the studies. No attempt has been made to supplement these studies with additional data from other sources.

8.4.2   Calcutta (India)

The setting
The Calcutta Metropolitan Area is in the state of West Bengal in the eastern part of India. Situated on the Ganga delta, at about 70 km north of the Bay of Bengal, it comprises a flat landscape with an average elevation of 6 m above sea level, and a maximum height of 9 m above sea level. The climate is tropical (hot and humid), with annual rainfall of about 1,600 mm, most of which falls in the monsoon months of July–September.

In the 1991 census, the metropolitan area had a population of 11.9 million. This represents a fivefold growth in the previous 70 years. Based on the 1991 census, 93 per cent of the population is classified as urban. About half of the population live in three municipal areas, the largest being Calcutta with 4.3 million inhabitants. Calcutta has a high population density, averaging 23,000

Table 8.5   Population characteristics of Calcutta and Howrah, 1981 and 1991

	Calcutta		Howrah	
	1981	1991	1981	1991
Area (km^2)	104.0	187.3	51.7	61.5
Population	3,288,148	4,385,176	744,429	950,435
Population density (persons per km^2)	31,617	23,413	14,399	15,454
Mean annual population growth rate, 1981–91 (% per year)		3.33		2.77

Source: Mukherjee *et al.*, 1995

persons per square kilometre, and this is much higher in the slum areas (up to 150,000 persons per square kilometre) where 35 per cent of the population live. Calcutta city also has an influx of an additional 1.5 million persons who commute daily for employment. Table 8.5 compares the population numbers and densities of two main municipal corporations (Calcutta and Howrah). Table 8.6 shows the age and gender distribution for both municipalities.

Health situation
The infant mortality rate (IMR) per thousand in 1993–94 was significantly lower in Calcutta (51.5) than for the state of West Bengal (70) and India as a whole (80). It was also below the national target of under 60 for the year 2000 (and close to the WHO "Health for All" target of 50 for the year 2000). Howrah municipality has a similar IMR of 53. There is, however, a wide intra-urban variation as indicated by data from one minority community in Howrah municipality, which had an IMR of just over 100 in 1992. Geographically, within the nine boroughs of Calcutta, the IMR ranges from 14.6 to 207, although these differences cannot be clearly attributed to socio-economic differences, patterns of health care, or gaps in the birth and death registration systems.

Cause-specific mortality data show that respiratory and circulatory diseases are the most important causes of death (17 per cent and 14 per cent respectively) in Calcutta. This contrasts markedly with data for the period 1970–78, when gastro-intestinal diseases were the main cause of mortality, followed by respiratory diseases (26 per cent and 11 per cent respectively). Reasons for the marked reduction in gastro-intestinal diseases include more effective treatment with antibiotics and oral rehydration therapy.

Table 8.6 Age and sex distribution of the population in Calcutta and Howrah, 1991

Age group (years)	Calcutta			Howrah		
	Males (%)	Females (%)	Total (%)	Males (%)	Females (%)	Total (%)
0–4	5.94	7.94	6.83	8.48	10.62	9.43
5–14	17.66	22.14	19.65	21.73	25.46	23.38
15–44	56.03	52.35	54.39	52.32	49.01	50.85
45–69	14.48	11.18	13.02	12.56	9.15	11.05
60+	5.89	6.39	6.11	4.91	5.76	5.29

Source: Mukherjee *et al.*, 1995

Environmental improvements may have also contributed to these changes, although these aspects have not been explicitly addressed.

Routine morbidity data are not readily available for Calcutta. Snapshots of the city have been made through health surveys and specific studies, but these are not sufficient to provide a detailed morbidity profile. Records of admission to the Calcutta Medical College Hospital during 1990–93 indicate that 4.2 per cent of admissions were related to gastro-enteric infections, compared with 7.3 per cent for cerebrovascular accidents. The extent to which these data are representative of Calcutta as a whole is unknown. A health survey carried out in a Calcutta slum in 1990 (N = 9,125), however, showed that gastro-enteric diseases affected 26 per 1,000 persons in the population at that time. The second most prevalent disease was respiratory infection (21 per 1,000), while all other diseases combined affected 35 per 1,000 of the population surveyed. The field study report stated that non-communicable diseases (including hypertension, myocardial infarction, diabetes mellitus, cancer, road accidents, bronchial asthma, alcohol and drug abuses and mental disorders) are emerging as significant health problems, although detailed data are not available to support these observations. Current data on road traffic accidents indicate that the rate is 32 per 1,000 in males and 14 per 1,000 in females, 31 per cent occurring in the 5–14 age group.

Local environment
Coal is the principal industrial fuel used in many industries in the Calcutta area, and thus acts as an important source of air pollution. Coal is also an important domestic fuel, and it has been estimated that 90 per cent of slum dwellings rely on coal and charcoal for fuel. Other air pollutants are the result of motor vehicle emissions (over half a million vehicles were registered in

1994). Many of these vehicles are old and inefficient in terms of combustion, and the problem is aggravated by traffic congestion. Currently there are nine air quality monitoring stations in Calcutta. Based on the data that these provide, suspended particulates can be seen as the most immediate air pollution problem, with mean concentrations of around 400 mg m^{-3} for some sites in recent years. However, sulphur dioxide levels are not high due to the low sulphur content of the local coal. Many small-scale industries also exist in the area, which when combined are likely to have a major impact on air (and also soil and water) pollution. Data are not available, however, to make estimates of their impact.

The Hooghly River is the main source of surface water for the city of Calcutta, with about $1{,}000 \times 10^6$ litres extracted daily. In addition, a further 500×10^6 litres are extracted from groundwaters. There are eight water treatment plants for surface waters, whereas groundwater is disinfected separately. In spite of these efforts, a survey carried out in 1992 showed that, of 945 water samples taken within the Calcutta area, 29 per cent were found to be "bacteriologically unsatisfactory". An additional problem is the intermittent water supply system, which necessitates storage of water by users. This is a potential source of water contamination and provides breeding places for the mosquitoes that spread malaria. Water scarcity is an important problem in some slum areas and squatter settlements along the canal banks.

Partially treated and untreated wastewater is a serious problem because of the occasional contamination of the municipal water supply and its use in irrigation of vegetables for human consumption. Open stormwater drains are often used for defecation and dumping of waste. Faecal contamination of drinking water occurs at source, in its transmissions through the public distribution system, and within the home through contamination of storage containers.

Inadequate collection and disposal facilities for solid wastes result in the accumulation of wastes on footpaths, roads and community dump sites. In addition, little separation of wastes occurs, so that household, hospital and industrial wastes are often stored, transported and disposed of together.

Food contamination takes many forms. Raw sewage and drainage water are used for irrigation of croplands and for nearby fisheries. Heavy metals have been identified in plants and fish fed with raw sewage. Parasitic and other microbial infections from these vegetables are a known public health risk.

Decision-making process
Several health indicators are being collected at the city level, including crude death rates, infant mortality rates, maternal mortality rates, percentage of newborn weighing at least 2.5 kg at birth, percentage of the population covered by primary health care, percentage of the population fully immunised, prevalence rate of malaria, and adult literacy rates. The quality of these

Table 8.7   Water pollution in Calcutta

Stage	Factors
Driving forces	Demand much larger than supply
Pressures	Contamination of drinking water due to leakage in sewage Wastage of drinking water through 8,000 stand points and public taps Uneven water distribution Domestic water storage
State	Faecal coliform at stand points have been identified Sporadic chemical pollution (arsenic, mercury, cadmium, chromium, lead) has been identified Vector breeding (*Anopheles*)
Exposure	Population not getting water supply is 30%
Effect	Gastro-enteritic diseases (morbidity and mortality) Occasional episodes of cholera and malaria
Actions	Diagnoses and treatment at government hospitals and public health centres Arrangement for mobile dispensary to specific groups in need Mass-awareness programmes by media coverage

data has not been assessed, but malaria prevalence for example is believed to be greatly underestimated.

Currently there is no systematic use of indicators for the purpose of decision-making. Often, decisions are reached after "much public suffering and furore". Constraints to environmental health action are seen to be the lack of public knowledge and awareness related to environmental hazards and their impact on health, lack of interdepartmental co-operation within and outside the government sectors, lack of political motivation, and lack of resources to fund relevant activities.

Comments and conclusions
During this study, some of the main problem areas identified were catalogued using the DPSEEA framework. This helped to identify data gaps as described in the example of water pollution (Table 8.7). While many health and environment indicators are currently being collected, their quality is unknown and they are often insufficient for the purpose of data linkage. Of the indicators listed in Box 8.2, the following are in use in Calcutta:
- Infant mortality rate (per 1,000 live births).
- Maternal mortality rate (per 1,000 live births).

- Percentage of the population with access to sufficient quantity of safe drinking water.
- Percentage of newborn weighing at least 2,500 g at birth.
- Percentage population covered by primary health care.
- Percentage of the eligible population that have been fully immunised according to national immunisation policies.
- Prevalence of malaria.
- Adult literacy rate.

Detailed information on these indicators was not provided. The authors, however, commented that these were "less than satisfactory and complete" in terms of data quality. This is an important issue, which is discussed further in section 8.5.

8.4.3   Cape Town (South Africa)

The setting
Cape Town is located on a peninsula at the southern tip of South Africa, within the province of the Western Cape. Government reports from 1993–94 have estimated the population of the Cape Town metropolitan area at between 2.5 and 2.9 million persons. For the districts of the city, various estimates of population size are used by the different institutions, based on census data. Data from the 1991 census have been shown to be particularly inaccurate in their count of the number of black people in peri-urban areas. Also, because of very rapid urbanisation and growth of informal settlements, accurate estimates are not easy to obtain. Table 8.8 provides information on the age and gender composition of the Western Cape province, which the authors of the report assume broadly to apply to the Cape Town metropolitan area.

Health situation
Quantitative information on population morbidity is limited. Mortality data are readily available, but studies have shown that there are two major problems with these data: under-registration and misclassification of deaths. Between 10 and 16 per cent of deaths in Cape Town are classified as "symptoms and ill-defined conditions". Uncertainties in the population data may contribute to errors in the calculated rates.

Infant mortality rates in Cape Town in 1993, based on official statistics, showed clear racial inequalities, ranging from 13.8 (per 1,000) in whites, 18.3 in coloureds and 33.9 in blacks. Nearly 18 per cent of all infant deaths were classified as ill-defined or unknown (indicating inadequate completion of death certificates). Pneumonia and diarrhoea each account for around 10 per cent of all infant deaths.

Table 8.8   Age and gender structure of the population of Western Cape province and South Africa

	Population (10^3)	0–4 years		5–14 years		15–64 years		65+ years	
		M (%)	F (%)	M (%)	F (%)	M (%)	F (%)	M (%)	F (%)
Western Cape	3,567	4.6	5.2	11.0	10.0	31.1	33.5	2.0	2.6
South Africa	41,008	6.1	5.9	12.2	11.9	28.5	31.5	1.6	2.4

Source: Lewin, 1996

Table 8.9   Infant mortality rates by place of residence and ethnic group in Cape Town, 1992

Ethnic group	Infant mortality rate (per 10^3 live births, 95% confidence intervals in parentheses)			
	Urban	Informal	Rural	All[1]
Black	23.2 (10.0–49.4)	34.3 (30.7–38.6)	16.1 (7.4–32.9)	33.0 (29.5–36.8)
Coloured	14.5 (12.5–16.8)	22.9 (9.1–51.8)	32.9 (26.0–41.3)	17.6 (15.6–19.8)
White	11.5 (7.6–17.2)	–	4.8 (0.2–30.8)	9.7 (6.7–13.9)
All	14.2 (12.4–16.3)	34.1 (30.5–38.2)	28.2 (22.6–35.0)	22.0 (20.3–23.8)

[1]   Including mixed type settlements   Source: Lewin, 1996

In a study made with data collected from notified births and deaths and supplemented with mortality records from 1992, infant mortality rates showed significant variation by race and type of settlement, as shown in Table 8.9. It was noted in this study that birth data are likely to be of good quality, but death data are likely to be under-reported in poorer (rural and informal) areas. Routine disaggregation of such data for planning, resource allocation and evaluation purposes was recommended in that study. The differences observed between this study, based on 1992 data, and the official statistics, based on 1993 data, are likely to be the result of the different methods used to obtain these estimates.

General mortality rates for 1993 also had a high percentage of deaths classified as "symptoms, signs and ill-defined" (13 per cent). Malignant neoplasms accounted for 15 per cent, homicides for 11 per cent and motor vehicle accidents nearly 6 per cent of all deaths. In the 15–44 age group, homicides accounted for 38 per cent and motor vehicle accidents for 13 per cent of all deaths.

Table 8.10   Percentage of the population having access to drinking water in the Western Cape Province (by ethnic group) and South Africa (all groups)

	Western Cape					South Africa
	Black	Coloured	Indians	Whites	All	
Piped water						
Internal	37.3	82.6	100.0	100.0	81.4	39.4
Yard tap	31.7	14.2	0	0	11.9	19.7
Public	29.6	2.7	0	0	6.3	17.3
Other	1.4	0.5	0	0	0.5	23.6

Source: Lewin, 1996

Local environment

Local authority Environmental Health Departments routinely collect data on workload (e.g. number of inspections), and these provide a limited basis on which to assess environmental health conditions or health risks. Data are collected on access to basic facilities (water, sanitation, waste collection) in informal settlements. In addition, data are available from several surveys, e.g. the Basic Subsistence Facilities survey which attempts to make an assessment of basic needs and to identify the constraints in meeting those needs. These and other relevant data sets (e.g. access to basic facilities in Cape Town; a national household survey on living standards and development) are in the public domain, in electronic format (but not in a form which is easily converted for data analysis).

Data on water and sanitation are available at the provincial level. Access to adequate and safe drinking water has been defined by health authorities as drinking water which is available within the home or within 100 m from the point of use; is adequate according to demand; and has a quality safe for health at the point of use. Based on a 1992–93 survey, it was found that 95.6 per cent of the urban population, 81.6 per cent of marginal urban, and 79.3 per cent of the rural population in the province had access to safe water according to these criteria. Access to water varies in relation to ethnicity, as seen in Table 8.10 (based on a survey carried out in 1993). The proportion of homes with piped water appears to be much higher in the Western Cape province than in the country as a whole, although caution should be applied in interpreting these statistics because the countrywide estimate was not obtained as a weighted sample.

Ethnic inequalities are also evident in terms of access to basic sanitation, with blacks in the least favourable situation. Table 8.11 presents data on

Table 8.11   Percentage of households with flush toilets in the Western Cape and South Africa, by ethnic group

Region	Black	Coloured	Indian	White	All
Western Cape	67.6	89.0	100.0	100.0	89.5
South Africa	32.4	88.0	99.6	99.8	52.1

Source: Lewin, 1996

households with flush toilets. Note that while caution should be applied in comparing Western Cape with the whole of the country (as in Table 8.10), important differences do seem to exist, with the province showing much better access among the black population. A recent survey, reported in 1994, found that for the Western Cape province 91.2 per cent of the urban population and 59.5 per cent of the peri-urban population has access to effective domestic waste removal systems.

Air quality is measured in a few places in the city (two sites for NO_x and ten sites for SO_2 and Pb). Some data are published in daily newspapers. These data are also available for research (e.g. currently, analysis is being done for the Cape Town Brown Haze study). Nevertheless, the monitoring network is not extensive enough to represent the exposed population accurately. In particular, monitoring stations are sparse or non-existent in areas of informal settlements. Because of the extensive use of wood, paraffin and charcoal fired appliances, indoor air pollution is likely to be a greater risk than outdoor air pollution in Cape Town, but data on indoor air pollution are not routinely collected.

Decision-making process
The current role of the city's Environmental Health Department is to identify problem areas (in relation to water, sanitation and shelter) and to bring these to the attention of the department responsible for the provision of the services in question. Similarly, regarding air quality, local authorities cannot initiate action (such as legal warnings or fines) unless a specific point source is identified. Local authorities have control at an early stage, because they have to approve the installation of industrial fuel burning appliances. There is no institutional link between the environmental conditions monitored and the provision of related services. Collaboration between the health sector and other sectors is mostly based on informal contacts.

Current environmental health debates in Cape Town are linked to broader national discussions in the health sector. One issue is the decentralised decision-making conditional on national policy to focus on the needs of the majority of the population. Another issue under debate is how to involve groups with little past involvement in decision-making in the policy-making process.

Overall, the usefulness of routine health and environmental data for decision-making is limited because data are of poor quality, scanty and rarely validated. While data linkage has been carried out as part of research projects, the results have not been fed to the decision-making process.

During the field studies, several interviews were carried out with key people involved in the field, and a workshop dealing with environmental health indicators was organised. These raised several important issues in Cape Town; for example, it was suggested that, locally, data collection needs to be related to programme objectives. Currently, it is not possible to assess the success of interventions, or to ascertain if environmental health conditions are improving. Also, there is no knowledge of the baseline conditions against which to measure improvements. The issue of data quality and presentation was also discussed in these meetings. It was pointed out that the quality of current data has not been examined closely, and that these data are not being presented in a way that is useful, relevant or accessible for decision-making.

Comments and conclusions

While consultations made for this study agreed that a core set of environmental health indicators needs to be developed for the city, attempts to do so were not successful. One important concern is that large volumes of data are collected locally for the national agencies and that these data have no direct relevance at the local level. Moreover, no feedback on what is collected at the local level is given once it is passed on to the national level. For the development of environmental health indicators it is crucial that these have local value, and are not seen merely as an additional burden on the system.

The main points discussed and agreed in the consultations were:

- The need to develop a rigorous and agreed set of environmental health indicators for the city, which are quantifiable and related to programme targets and objectives.
- A need to improve data accessibility and relevance at all levels.
- An emphasis on "quality" needs to be built into the whole system of environmental health.
- The need for feedback between communities, the environmental health system and between the different levels of management structures.
- A need for capacity-building in environmental health at all levels.

Of the proposed list of indicators (Box 8.2) several were identified as being used locally, but no further information was obtained regarding their quality. Indicators were not considered within the DPSEEA framework. The indicators used were:

- Infant mortality rate (per 1,000 live births).
- Maternal mortality rate (per 1,000 live births).

- Percentage of the population with access to sufficient quantity of safe drinking water.
- Percentage of the population with access to hygienic excreta disposal facilities.
- Percentage of people without access to adequate food supply.
- Percentage of newborn weighing at least 2,500 g at birth.
- Percentage population covered by primary health care.
- Percentage of the eligible population that have been fully immunised according to national immunisation policies.
- Adult literacy rate.
- Percentage GNP spent on health.
- Percentage of national health expenditure devoted to primary health care, health centres and regional hospitals.

8.4.4   Cotonou (Benin)

The setting
Cotonou is the commercial capital of Benin, and stands between the Nokoue lake in the north and the Atlantic coast in the south. In 1994, the population of Cotonou was estimated at 580,000 (4.9 million for Benin as a whole). The city has shown a sevenfold increase in population since the 1961 population census, when it had about 78,000 persons. This growth is attributed to natural growth and migration. Cotonou has a diverse ethnic composition. Major ethnic groups are the Fon (42 per cent), the Adja (16 per cent), and the Yoruba (12 per cent). More than half of the people are Animists, following indigenous religious beliefs and practices.

From 1972 to 1990 Cotonou was divided into six administrative districts. These were abolished after 1990 and now there is just one administrative zone. The national health system, however, still functions on the basis of the former six districts (numbered I–VI).

Annual rainfall in the south of Benin, where Cotonou is located, is about 1,200 mm. There are two rainy seasons, the main one occurring from April to July, when the city is regularly affected by floods. The coastline is seriously affected by erosion and is retreating at a rate of about 17 m a^{-1}, apparently due largely to disturbance to the coastal sediment system, caused by construction of the city's port.

Health situation
The local authorities are required to record all deaths, because this is a condition for burial. Nevertheless, many deaths remain unregistered either for economical or cultural reasons. Morbidity data are also available from the main hospitals, while some infectious diseases are notified to the health

Table 8.12   Incidence of the five most prominent diseases in Cotonou and Benin, 1994

| | Incidence (per 10^3) | |
Disease	Cotonou	Benin
Malaria	110	340
Upper respiratory	30	44
Anaemia	20	13
Diarrhoeal	18	27
Injuries	11	16

Source: Soton *et al.*, 1995

department by public health centres. In addition, however, there are many private clinics which are not integrated into the national health system, and morbidity data from these are not routinely collected.

Data from 1994 for Cotonou and for Benin on the five most predominant diseases are given in Table 8.12. Overall, Cotonou appears to have a lower incidence for most of these diseases than the national average. Within Cotonou, however, there are marked geographical and social variations in health status. Health zones I and IV are the most deprived areas in the city with 37,000 and 63,000 inhabitants respectively (close to 20 per cent of the population in 1992). The main problems in these zones include poor housing, lack of clean water and risks of flooding. These two zones also have the highest observed incidence of malaria, upper respiratory conditions, anaemia and diarrhoeal diseases.

Children are especially at risk from several of these diseases (Table 8.13). In all cases, infants under one year of age are seen to be those most at risk, with rates between 2.5 and 3.5 times those in the 1–4 year age group, and four to five times higher than the average for the whole population. Rates of malaria are especially high, with an incidence of 450 per 1,000 among children under one year of age.

Local environment

Basic data on service provision (e.g. access to water facilities and waste removal facilities) are collected by government agencies. No specific information on environmental pollution is available. Air pollution is a recent and increasing problem in Cotonou because of the increasing fleet of old cars and growth in the total number of motorcycles, but there is no monitoring system for air quality in the city. For the more traditional environmental problems, such as access to clean water and sanitation, data are also sparse. The only

Table 8.13   Incidence of main diseases in young children in Cotonou, 1994

| Age group | Incidence (per 10^3) | | | |
	Malaria	Upper respiratory tract infections	Diarrhoea	Anaemia
Under 1 year	450	176	93	81
1–4 years	150	50	26	34

Source: Soton *et al.*, 1995

available information on access to drinking water, for example, was a report which estimated that 88 per cent of households in Cotonou obtain water from unprotected wells. The report stated that sanitation and methods of waste disposal were inadequate to meet current needs, but no data were provided to help quantify the problem.

Decision-making process
There is no formal process for decision-making in environmental health in place in Benin. Problems are identified by residents, the mass media or in research or government reports. The problems identified are brought to the attention of the relevant government authorities. At times, international organisations and NGOs become involved. Lack of resources is often the main constraint on attempts to implement action.

Geographical Information Systems software has been used as an analysis tool to aid the decision-making process, as in the case of the monitoring of dracunculiasis. Innovative approaches to solve immediate problems have also been tested, for example to deal with the limited extent of waste removal services. This was approached by an NGO through a local project, aimed both at removing waste and at creating employment among young people.

Comments and conclusions
Information is available for several of the indicators listed in Box 8.2. However, these are based not on specific, local data, but by extrapolation from national information:
- Life expectancy at birth.
- Infant mortality rate (per 1,000 live births).
- Maternal mortality rate (per 1,000 live births).
- Percentage of the population with access to sufficient quantity of safe drinking water.
- Percentage of the population with access to hygienic excreta disposal facilities.

- Percentage of people served by public waste removal service.
- Percentage of newborn weighing at least 2,500 g at birth.
- Percentage population covered by primary health care.
- Percentage of the eligible population that have been fully immunised according to national immunisation policies.
- Prevalence of malaria.
- Adult literacy rate.
- Percentage GNP spent on health.

Overall, it is evident that only very basic information regarding health and environment currently exists in Cotonou. This scarcity of data would appear to inhibit effective decision-making and impair attempts to improve the environmental health situation. Capacity building activities, aimed at improving data collection and analysis, are thus an important first requirement for the city.

8.4.5 Managua (Nicaragua)

The setting
Managua, the capital city of Nicaragua, is located at the shore of lake Xolotlán, within the Pacific region of the country. The city is situated in an area of seismic activity and has experienced two important earthquakes this century, the latest in 1972 when the entire city centre was destroyed and more than 10,000 casualties occurred. Managua had just over one million people in 1994. This represents 24 per cent of the country's population and nearly 85 per cent of the population of the department (also named Managua). Annual urban growth is 3.2 per cent, but during the 1980s the population increased rapidly, primarily due to rural–urban migration as a result of military action. There is a large shortage of housing units, with the result that there are some 270 informal settlements in the city. The official (registered) number of housing units is 118,000, with an average of 7.6 persons per house: overcrowding is observed in 50 per cent of these houses. The average population density is 100 inhabitants per km^2 and about 22 per cent of the population is younger than 15 years.

Health situation
Life expectancy for the period 1990–95 was estimated at 66 years. Until very recently, morbidity data collected by health centres and hospitals have been of good quality. The introduction of user fees and the increase in private health care provision, however, have made data collection more difficult. The quality of mortality data is satisfactory but under-reporting of infant deaths is a recognised problem. Mortality data collected include age, gender, place of occurrence, main and associated causes of death.

Table 8.14   Leading causes of morbidity in Managua, 1994

Disease	Morbidity (rate per 10^5)
Acute respiratory infections	21,587
Diarrhoeal diseases	5,514
Cholera	1,013
Malaria	896.2
Dengue fever	647.3
Injury by animal suspected of rabies	410.2

Source: Gonzalez *et al.*, 1995

The health situation in Managua was described by this study as one of *"epidemiological polarisation"*, where *"the traditional diseases of poverty continue and new health problems related to rapid industrialisation and modernisation increase"* (Gonzalez *et al.*, 1995). Vector-transmitted diseases such as malaria, dengue fever and leptospirosis have been defined as priority health problems in the city. Acute respiratory infections, cholera and acute diarrhoeal diseases are all endemic. The disease surveillance unit in the Ministry of Health publishes a monthly epidemiological report of key communicable diseases. Data are assembled from hospital and health centre records. The main health indicators for 1994 are given in Table 8.14.

Occupational health and safety is deficient. Studies have shown that 30 per cent of formal sector workers in Managua are exposed to physical and/or chemical contaminants, and an annual accident rate of 12 accidents per 1,000 workers employed in the formal sector has been reported (2 per cent of these are fatal). Only 15 per cent of the economically active population is covered by the social security system.

Local environment
The environmental problems likely to affect the population's health are many. During a workshop organised as part of this study, participants representing different sectors and research institutions attempted to define the major environmental health problems in the city. Eight important problem areas were identified:
- The decline in quantity and quality of sources for drinking water.
- Increased urban growth and urban poverty.
- An increasing deficiency (quantity and quality) of housing.
- Increased generation, and inadequate collection and disposal, of solid waste.
- Contamination of soil, air and water sources by specific chemicals and metals.

- Insufficient coverage and inadequate disposal of domestic and industrial wastewater.
- High morbidity and mortality rates resulting from waterborne, vector-transmitted diseases and by occupational risks.
- Lack of legislation and educational programmes in environmental health.

Decision-making process

Lack of co-ordination between sectors was seen as a major impediment to environmental health management, contributing to the limited impact of most interventions. Attention to environmental health problems is often the result of pressure from community groups, when situations reach critical levels. There is no systematic action to approach existing problems. Whenever interventions are implemented, they tend to be only palliative, which in the long term may produce a worsening of environmental health problems. There is also a lack of technical capacity. Capacity building in the areas of statistical methods, epidemiology and environmental health are thus important priorities.

Comments and conclusions

A workshop on Environmental Health Indicators for Decision-Making was organised as part of the study. This was important in bringing the main interested parties together to identify priority areas and in improving understanding about the environmental health situation in Managua. Unfortunately, the views of community representatives were not presented at the workshop.

The list of indicators provided in Box 8.2 was not directly considered in the study, but the workshop proposed a preliminary set of environmental health indicators for decision-making, which would require further testing and development as part of a consultative process. These covered several areas including poverty level, urban growth, industrial pollution, water coverage and wastewater re-use. It should be noted that some of the proposed indicators are, more strictly, "problem areas", for which several more specific indicators would need to be defined for monitoring purposes. Because the city is linked to the Healthy Cities project (Managua Municipio Saludable), it would also be useful and timely to develop environmental health indicators as part of other Healthy Cities projects.

It was evident throughout this study that systematic training and sensitisation of future managers and key personnel in the area of environmental health were important needs. Addressing poverty, health and the environment in an integrated manner and with a participatory approach was seen as the most appropriate and effective model for action and change. This was shown in a pilot project aimed at building a "healthy neighbourhood", implemented by an environmental organisation in a poor locality in Managua. The study also illustrated the need for political commitment if information gathering and use is to lead to improved health status. During the 1980s, a joint programme was undertaken by the Ministry of Health, Ministry of Labour and trade unions aimed at promoting "healthy workplaces". As part of this, workers' exposures and health were monitored on a routine basis. The effectiveness of this approach was demonstrated in one specific case, where evidence of lead poisoning of workers in a battery factory led to the temporary closure and renovation of the factory. In recent years, however, this programme has lost priority and has practically stopped. Routine monitoring is no longer carried out and workers' health is currently monitored only through *ad hoc* studies performed by research institutes.

8.4.6   Manila (The Philippines)

The setting
Metropolitan Manila (the National Capital Region) is located in Luzon, the largest island in the Philippines. Manila is composed of eight cities and nine municipalities. Its population in 1992 was about 8.4 million, representing 13 per cent of the population of the country as a whole. About 40 per cent of the population were classified as "urban poor", half of whom live in urban slums. Some 12 per cent of the population is under five years old, and one third are under 15 years of age. The inhabitants are wealthier than in the rest of the country, with twice the average income and more than double the average savings. However, the unemployment rate is high (27 per cent in 1995). In 1993 there were nearly 36,000 establishments employing five or more persons (a total of 1.35 million persons). This represented half of the total registered labour force in the country. The informal labour sector in Manila is estimated at 1.9 million. These include a variety of activities, such as street vendors, repair shops and small-scale cottage industries.

Health situation
There has been a dramatic improvement in health in Manila in recent years. Infant mortality was 19.2 per 1,000 live births in 1994, compared with 45.8 per 1,000 only four years earlier. Life expectancy at birth, at 67 years, is the highest in the country. Pneumonia, cancer and tuberculosis were among the leading causes of mortality in 1989–93 (Table 8.15). Among the main causes of morbidity are respiratory diseases such as pneumonia and bronchitis, and waterborne diseases such as diarrhoea (Table 8.15).

Routine health data are available from mortality registries and from facility-based morbidity records. The quality of these data is known to be deficient, however, and under-registration (particularly of infant deaths) is recognised as a problem in the Philippines, although less so in Manila.

Table 8.15 Leading causes of mortality and morbidity in Metro Manila, 1989–93

Disease	Mortality (rate per 10^5)	Disease	Morbidity (rate per 10^5)
Pneumonia	70.1	Diarrhoea	928.3
Cancer	39.1	Bronchitis	887.4
TB (all forms)	38.1	Pneumonia	514.8
Vascular disease	36.2	TB (all forms)	247.2
Accident	28.1	Influenza	206.2
Hypertensive disease	21.6	Heart disease	63.4
Septicaemia	10.7	Measles	62.7
Liver disease and cirrhosis	9.7	Chicken pox	57.9
Diabetes	9.3	Hepatitis (viral)	51.6
Kidney disease	8.7	Cancer	21.0

Source: Torres and Subida, 1996

Local environment

Air pollution is one of the major environmental problems in Manila. This derives from both industrial and mobile sources. Between 1990 and 1994 the number of vehicles registered increased by 30 per cent to 960,000. Currently, the Manila area accounts for about 40 per cent of the country's fuel consumption, and nearly 50 per cent of all industrial establishments in the country. A national occupational health survey of 3,426 establishments in 1991 showed a high proportion of workers experiencing occupational health hazards. It was found that 96.8 per cent of the workplaces surveyed had significant exposures to suspended particulates and 75.6 per cent to excessive noise levels. Other problems identified included exposure to dust and gases, abnormal working positions and repetitive physical tasks.

High levels of atmospheric lead were an important problem before 1994, when low-lead and unleaded petrol were not available. A study of blood-lead levels in children aged 6–14 was undertaken in 1993. This showed significantly higher levels in street vendors than in school children, and much higher levels in boys than girls (Table 8.16). Since then, however, levels of atmospheric lead exposures are believed to have been reduced as a result of the introduction of unleaded petrol.

Epidemiological studies have demonstrated the links between air pollution and respiratory conditions in the city. In a study of jeepney (i.e. open vehicle) drivers, it was shown that several respiratory symptoms were more prevalent among this group than among drivers of air-conditioned buses (Table 8.17).

Table 8.16   Distribution of blood lead levels in children aged 6–14 years in Metro
Manila, 1993

School children	No. of subjects	% with levels >20 mg/100 ml	Child vendors	No. of subjects	% with levels >20 mg/100 ml
Boys	210	15.2	Boys	80	35.0
Girls	177	4.5	Girls	21	23.8
All	387	10.3	All	101	32.7

Source: Torres and Subida, 1996

Table 8.17   Prevalence of chronic respiratory symptoms among jeepney and
air-conditioned bus drivers in Manila, 1990

Symptoms	Prevalence (%)	
	Jeepney drivers (n = 314)	Air-conditioned bus drivers (n = 262)
Chronic cough	19.1	8.8
Chronic phlegm	26.4	11.1
Wheezing	16.6	10.3
Shortness of breath	24.5	15.3

Source: Torres and Subida, 1996

On the evidence of 1992 data, solid wastes continue to be a problem with about 70 per cent of domestic and industrial wastes being collected and deposited in landfills or open dumpsites. The remainder is disposed of informally; for example by private hauliers, by recycling, by burning, or by dumping in waterways or vacant land areas. People who live on or near dumpsites scavenge for a living. The largest dumpsite in Manila, known as the "Smokey Mountain" dumpsite, was home to some 3,000 families (approximately 21,000 persons) in 1992. Prevalence of malnutrition (underweight and stunting) amongst these families was almost 20 per cent, and one in ten children were found to be anaemic, and one third suffering from multiple parasitism. Because of the continuous burning of wastes, the mean 24 hour exposure of child scavengers to total suspended particles was 4,600 mg m^{-3} nearly 30 times the national guideline.

All eight major river systems in Manila are considered virtually dead because of the organic loading from industrial and domestic waste discharges. These rivers drain towards the Laguna Lake, the largest in the South East Asian region. Plans for the lake to be the drinking water source for

Manila were rejected in 1990. In spite of this, the 1992 national health survey showed that provision of water supply is high at 96 per cent (82 per cent of households having household connections, the rest relying on other means including public water taps). Moreover, 85 per cent of households have access to sanitary facilities.

Decision-making process
The Department of Health and the Department of the Environment and Natural Resources are mandated to address, respectively, the health and the environmental concerns of the country. With the adoption of Agenda 21, steps have been taken to incorporate health and environment issues within development projects. In its National Health Plan, for example, the Department of Health identified the following needs:

- To define basic minimum health needs of each citizen and to propose ways to meet these within specific environmental and cultural contexts.
- To set national standards to control communicable diseases.
- To improve information systems to identify vulnerable groups.
- To advocate enforcement of laws and regulations particularly on health impact assessment.
- To develop technical expertise in assessing health risks caused by development projects.
- To seek participation of other government bodies with mandates to oversee health, environment and development issues.

These strategies have been reinforced by the creation of the Interagency Committee of Environmental Health in 1987, and the Philippine Council for Sustainable Development in 1992. The latter has mandated the integration of health impact assessment within the current environmental impact assessment system of the country.

In 1992, a devolution process of certain specific functions from central to local government began. This has at times resulted in unclear delineation of responsibilities, particularly in the implementation and regulation of environmental health policies and guidelines. An additional constraint has been the limited resources in terms of skills and technical capabilities at the local level, needed to address the multifaceted nature of environmental health responsibilities.

The field study identified three major groups involved in the decision-making process:

- The government sector, including national and local level agencies and the legislative bodies, which are responsible for the provision of services and setting policies and standards.
- The private sector, which includes industry and business sectors.
- The public, who are ultimately affected by pollution.

Decision-making occurs in two ways: by the use of scientific data (indicators) on health and environment and by pressures set by advocacy groups such as community organisations and the media. For example, the reduction of lead in petrol and the drafting of the Clean/Healthy Air Act resulted from an initial set of studies dealing with environmental emissions and epidemiological investigations. Two Senate committees reviewed the pollution control laws and regulations and problems related to their enforcement, the programmes of the different agencies involved and the scientific data available. The review included both technical working group investigations and public hearings with the participation of the national government agencies, the oil companies and the general public.

Comments and conclusions
In Manila several epidemiological studies have been conducted and scientific information has been used successfully to aid the decision-making process, as in the example given above. Data linkage has been performed in special projects conducted by consultants for government agencies, and also by academic and research institutions. An important study was the Philippine Environmental Health Assessment study which made an assessment of the impact of environmental pollution on human health. That study concluded that data quality must be improved for it to be used for decision-making.

One of the main problems in using routine data, however, is ensuring their correct interpretation. It has been shown in the Philippines that those regions most affected by certain forms of environmental pollution (particularly urban pollution) are also the most developed in terms of prosperity and access to services and facilities. The health impact of pollution therefore tends to be masked by these social and developmental factors.

Environmental health problems identified in the study were classified according to the DPSEEA framework. One example, urban housing, is given in Table 8.18. This example is especially pertinent because of the concentration of industry in the city, and its link with other environmental health problems.

The following indicators from Box 8.2 are in use in Manila, although all are subject to the quality constraints discussed above:
- Life expectancy at birth.
- Infant mortality rate (per 1,000 live births).
- Maternal mortality rate (per 1,000 live births).
- Percentage of the population with access to sufficient quantity of (safe drinking) water (NB: Quality not assessed).
- Percentage of the population with access to (hygienic) excreta disposal facilities (NB: Hygienic aspect not assessed).
- Percentage of people exposed to high concentrations of health damaging air pollution (outdoors only).

Table 8.18   Urban housing problems in Manila described using the DPSEEA framework

Stage	Descriptive indicator	Action indicator
Driving forces	Industrial development Workforce demand	Development of other industrial areas outside Metro Manila
Pressures	Number of industries Size of squatter population	Dispersion to other parts of the country Building low-cost housing units
State	Level of social and health services State of housing structures	Coverage of water supply and sanitation facilities Relocation of squatter population
Exposure	Water supply and sanitation levels Crowding index	Provision of housing units and facilities
Effect	Number of diarrhoeal diseases Malnutrition Respiratory diseases	Treatment Immunisation Prevention measures

- Percentage of newborn weighing at least 2,500 g at birth.
- Percentage of the eligible population that have been fully immunised according to national immunisation policies.
- Prevalence of malaria.
- Incidence of acute enteric infections.
- Adult literacy rate.
- Percentage GNP spent on health.

8.4.7   Talcahuano (Chile)

The setting

Talcahuano is an industrial city of nearly 250,000 inhabitants, with serious and diverse environmental health problems. It is located on the coast of Chile's 8th Region, Bio Bio, which has a population of 1.7 million (approximately 13 per cent of the country's population, the third most populated of thirteen regions). The city of Talcahuano comprises one of 49 municipalities which form the region. The city is linked with the city of Concepción, which together constitute the metropolitan area of Concepción-Talcahuano with a combined population of about 580,000 persons.

Talcahuano has been affected by pollution for many years from large emissions from industrial processes, mobile sources and uncontrolled decay of organic matter from the local fishing industries. Recognising these problems, the government set up a commission in 1992 to develop an Environmental Restoration Plan for Talcahuano. The general objectives were:

- To improve the health and quality of life of the people.

Table 8.19   Mortality rates for Talcahuano, Concepción, the 8th Region and Chile as a whole, 1993

	Talcahuano	Concepción	8th Region	Chile
Population	248,543	331,027	1,734,305	13,348,401
Crude mortality rate (per 10^3)	4.4	5.3	5.8	5.5
Infant mortality rate (per 10^3)	12.2	14.7	15.5	13.1

Source: Salinas *et al.*, 1995

- To preserve and restore natural resources.
- To make industrial development compatible with urban activities.

Health situation
Annual mortality data are available from the National Institute of Statistics, with a delay of 18 months. These data can be purchased in electronic format and include the following variables: gender, marital status, age, educational level, employment status at death, occupational activity, date of death, municipality of residence, urban or rural category, cause of death, type of medical certification and birth date of the decedent. Data on infant mortality include weight at birth, gestation age, nutritional status, marital status of mother, parity, number of decedent children, date of last delivery and educational level. Information on the cause of death is known to be accurate (above 85 per cent). Mortality rates are given in Table 8.19.

Morbidity data are available from public primary health care centres and hospital discharge statistics. Unfortunately, data from primary health care centres are not representative of the population base, because of difficulties of access (e.g. distance) and the use of the private sector (which is estimated at around 30 per cent). In addition, data quality is not optimal because of frequent lack of diagnosis in the medical records. By contrast, data on hospital discharges are increasingly being recorded electronically, which will enhance their use for data linkage studies.

Local environment
The field study identified several environmental health problems that affect the quality of life in Talcahuano. The main problems are air pollution from chemical, petrochemical, steel, food and fish industries around the city, plus the many small scale industries, all in proximity to the urban areas. The problem is aggravated by mobile sources, which include large vehicles servicing industry and the harbour. Estimated atmospheric emissions are given in Table 8.20.

Table 8.20   Estimated emissions of air pollutants in Talcahuano, by source, 1992

Activity	Emissions (t a⁻¹)		
	SO_2	NO_x	Particulate matter
Industry	11,520	4,372	713
Transport	114	602	124
Domestic activities	19	148	485
Bakeries	1	10	42

Source: Salinas et al., 1995

Urban activities generate a monthly mean of 4,500 tonnes of waste. Solid industrial waste, from more than 600 productive activities, has not been quantified. There are ten unauthorised dump sites, which are periodically cleaned by the local municipalities. Illegal disposal of solid wastes and inadequate municipal waste landfills are a threat to groundwaters. While over 85 per cent of households are linked to the city's sewage system, the functional capacity of this system is constrained by poor maintenance and discharge of residuals from some industries.

According to a study conducted by the local municipality, the San Vicente harbour, in the west of the city, receives nearly 270,000 m³ of liquid wastes daily, mainly from industrial processes. Talcahuano harbour to the north is similarly affected, receiving about 80,000 m³ per day. The Bio Bio river, in the south of the city, receives 40 per cent of the city's wastewaters (26,000 m³ per day) and 230,000 m³ of industrial liquid residuals from an oil refinery.

Decision-making process
Since 1990 there has been considerable concern, nationally, regarding the state of the environment, and in 1994 the parliament approved the Basic Law of the Environment which established the tools for environmental management, the principle of liability of environmental damage, and the institutional framework to implement regulations. The law also promotes education and research and lays down procedures for setting emission and environmental quality standards. It defines areas of serious contamination, requiring specific intervention, and areas of borderline environmental levels, requiring prevention measures. The law also incorporates community participation into the process of evaluation, planning and standard setting.

The decision-making process does not use indicators in any systematic way. Decisions are often taken as a response to urgent situations. Current laws dictate that the municipalities must contribute to the protection of the environment and the health of the people. Municipal actions in this area have consisted of commissioning research studies, such as an air pollution survey, the results from which were used to make recommendations for improvements in industry.

Community involvement is hampered by the low degree of commitment of the inhabitants. An earlier study on this issue indicated that a large proportion of the residents of Talcahuano live there because there is no alternative for them, and would migrate if they could. The presence of industry is an important source of employment in the region. This community perception tends to impede the creation of strong community associations to lobby for environmental improvement.

Comments and conclusions
As part of the field study, several important stakeholders were brought together at a workshop to discuss the environmental health situation of the city. Data linkage was not performed due to a lack of more detailed local data, but maps of selected health and environment indicators (related to air pollution), aggregated by municipality for the region of which Talcahuano forms part, were prepared. The study also developed proformas for registering medical attendances for specific diagnoses, with predetermined categories to minimise the problem of lack of diagnosis or unreadable handwriting in medical records.

Of the indicators given in Box 8.2, the following were identified as in use, although their quality has not been assessed:
- Life expectancy at birth.
- Infant mortality rate (per 1,000 live births).
- Maternal mortality rate (per 1,000 live births).
- Percentage of the population with access to sufficient quantity of safe drinking water.
- Percentage of the population with access to hygienic excreta disposal facilities.
- Percentage of people served by public waste removal service.
- Percentage population covered by primary health care.
- Percentage of the eligible population that have been fully immunised according to national immunisation policies.
- Adult literacy rate.

8.5   Field studies evaluation
This section evaluates the findings from the six field studies, as a basis for assessing the usefulness and effectiveness of the HEADLAMP process. The studies are considered in relation to three broad areas: problem identification and assessment, data availability and quality, and relevance to decision-making processes.

Table 8.21   Methods of problem identification used in field studies

	Literature review	Interviews	Workshop
Calcutta	Yes	No	No
Cape Town	Yes	Yes	Yes
Cotonou	Yes	Yes	No
Managua	Yes	No	Yes
Manila	Yes	_[1]	_[1]
Talcahuano	Yes	No	Yes

[1]   Manila benefited from a similar process undertaken prior to this study

8.5.1   Problem identification and assessment

Methods of problem identification

Three main methods were specified in the project guidelines for identifying environmental health problems in the study areas: a review of the literature (which included the identification of existing data and their sources), interviews with local experts and the organisation of workshops to bring key people or groups together to review the local environmental health situation. Table 8.21 shows the use of these methods in the six field study areas. As this indicates, not all methods were used in every case, although all employed a literature review, and in most cases a second approach was also used (see Table 8.21). Differences in approach to some extent reflected the level of development of the information structures and decision-making processes in the six field areas. Interviews and workshops tended to be more feasible to set up in those cities with better developed systems.

In all six cities there were sources of data which, even if not complete, were useful in indicating key environmental health issues, and in pointing the investigators towards areas where further consultation was required. Marked differences nevertheless existed in the quantity, extent and quality of the data available in the different study areas. Manila, for example, is comparatively rich in environmental health research, while Cotonou has a major information deficiency. Talcahuano is in the process of building up its information system where none existed, but currently it is still incomplete.

The methods followed, nevertheless, do not ensure that all existing environmental health problems were identified at each site. Indeed, the question of defining environmental health problems is a complex one, for any issue can be perceived from many different perspectives and specified in many different ways. This raises the further question as to whether secondary data are a sufficient basis from which to detect problems. A major difficulty in this context is that these data often only exist because a problem has already been defined. Problems thus tend to be seen from the perspective of those who have the ability and resources to collect and present "hard" information. New problems, for which data have not yet been collected, or problems experienced by groups who have little voice in the data gathering process, are thus likely to be unrepresented in secondary sources. Reliance wholly on secondary sources is thus likely to lead to strong biases in problem recognition.

For these reasons, the involvement of key stakeholders is essential, both to confirm the validity of any problems identified from the available secondary sources, and to help identify other problem areas to which these data do not refer. Interviews and workshops provide a useful way of obtaining these additional insights. It can thus be expected that problem definition was more complete and more balanced in those study areas where a combination of approaches was used. Nevertheless, these methods also contain biases; for example depending on the choice of participants, their own knowledge and experience, and their ability or willingness to express their views. In any such survey, therefore, it cannot be assumed that all issues of concern have been identified. One way of strengthening the process is to provide extensive feedback on the results to the communities concerned, and to invite further response and comment.

One common deficiency in all studies was that they focused mainly on the environmental exposures, health outcomes and possible solutions. Rarely was consideration given to the root of the problems, i.e. the driving forces and pressures, as defined in the DPSEEA framework. Again, this reflects both the focus of most secondary information, and the main experience and interests of most stakeholders; their concern tends to be targeted at the more immediate source and effects of any problem, rather than at more remote, underlying causes. For policy purposes, however, this is unfortunate, because it may encourage a short-term perspective, with an emphasis on remediation rather than avoidance. It may also inhibit the ability to recognise the causal links that exist between many problems, and thus the opportunity for more integrated control.

Ranking of environmental health problems
The ranking of environmental health problems in terms of their importance is important particularly in developing countries where limited resources are available, and decision-makers need to make an informed choice on their investments. It is, therefore, an essential part of the HEADLAMP process. In practice, however, ranking of environmental health issues in any rigorous or representative way is far from simple, as recent attempts to establish National Environmental Health Action Plans have shown (Briggs *et al.,* 1998). The

importance of any issue depends not just on the number of people affected, but on the clinical severity of the health effects, on the level of public fear about the risks involved, on the ability of people to control or cope with the effects, and on the costs of remediation or control. As noted above, people's perceptions of any environmental health problem are also likely to vary depending on their experience and circumstance. In order to assess problems, therefore, a wide range of stakeholders would need to be consulted, using methods such as Delphi techniques (Fink *et al.*, 1984; Richey *et al.*, 1985a,b; Jones and Hunter, 1995) in a systematic way. One good example of a successful ranking is the recent work in Sweden where a multidisciplinary group of experts estimated the number of cases attributable to various environmental agents, as part of a national plan to reduce environmental health risks in the country (Victorin *et al.*, 1997, 1998). This exercise also raised several issues such as how to account for environmental risks that are poorly investigated and how to deal with potential risks rather than less severe but existing risks.

No formal attempt to rank or weight issues was made in the field studies described here. An informal process was applied in Managua, based on consultation with local experts, but although this involved strong scientific representation, there was no participation from community groups. The findings are thus likely to be partial. For future studies, a two-stage approach to ranking might be proposed. This would start with a general consultation bringing together all relevant and interested parties, aimed at defining and broadly ranking issues of concern. In the second stage, a scientific assessment of these issues could be undertaken, including an analysis of existing data, and the collection and analysis of specially collected data, where required.

8.5.2   Data availability and quality

Data availability
What constitutes relevant environmental and health data clearly depends on the selection of environmental health problems. In general, data of relevance to the issues recognised were identified in these studies, where data existed. The extent to which this represented a state of total capture for the relevant data is, however, impossible to assess, because there is no independent inventory of the available information. Indeed, the lack of such metadata (data about data) is one of the main difficulties facing studies of this sort, and one of the main areas where improvements need to be made.

In several cases, lack of available data certainly presented a major constraint. The situation was most severe in Cotonou, where very little health and environment data were available, but all field studies reported important health and environment issues for which data were not collected. Where data were available, problems of data quality also tended to occur; these need to be addressed if the data are to be used with confidence.

Where data existed, the sources were identified. Some studies found, and were able to give, more specific details regarding the data sources and the format of these data. This latter aspect is important because it has implications regarding data use. For example, paper records are of limited use when not already aggregated and analysed. Electronic data format has other constraints related to cost and access. Issues of level of data aggregation, timeliness, frequency of updating and quality assurance and control are also important is assessing the usefulness of the data. As noted before, therefore, there is often a need to improve metadata relating to environmental health data at the local level.

Data quality
Data quality was not assessed for each separate indicator, but it was discussed in general terms in the field studies. In the case of health data, a number of issues of data quality were identified. In all sites, mortality data, even if deficient in some cases, were available. Morbidity data are harder to obtain because of the lack of a central register (as for mortality data). The emergence of computerised data collection methods is likely to change this situation considerably in the near future, as indicated by the Talcahuano study. Electronic data collection will make a major contribution by allowing data from different sources (e.g. different hospitals and clinics from one city) to be put together in a relatively short time.

Problems with the available health data included inconsistencies in diagnosis, incomplete reporting (especially for morbidity data) and uncertainties in the denominator data (e.g. total or age-stratified population). In many cases, local data were also aggregated only at a relatively coarse scale, making it impossible to detect local patterns of health outcome. A minor, yet important, problem in the case of paper records (and a source of error in data encoding) is the illegibility of many health records (as found on Talcahuano).

Data on human exposures were generally more problematic. In many instances, only general "pointers" of environmental problems, instead of actual indicators, were available. Air pollution data are some of the most widely available, and are usually based on the continuous and systematic monitoring of air quality at a network of sites. Nevertheless, even these data are often far from ideal as a basis for exposure assessment. In particular, monitoring sites are not always located where people live and, because people move within micro-environments, data on environmental concentration provide a relatively poor proxy for exposure. The range of pollutants monitored may also not reflect the main environmental health risks of concern (e.g. monitoring of fine particulates is limited in most cities). Data on other potential human exposures, such as quality of drinking water, access to sanitation, and type of shelter, are often of an even poorer quality both in terms of their completeness and their accuracy. Overall, it is evident that the quality of environmental data needs to be greatly improved if they are to form a reliable basis for environmental health decision-making. In addition, the data need to be available at a spatial resolution that enables problem areas to be identified, where specific interventions are most needed.

8.5.3   Relevance to decision-making

Local decision-making processes and responsibilities
The decision-making process at the local level was described in all studies. In most cases it is characterised by lack of use of routinely collected data and scientific research, and by being mostly reactive (to community pressures or to critical situations) rather than proactive and protective of health and the environment. In both the Philippines and Chile there are clear mandates to address health and environment concerns. In Manila and Talcahuano, however, community pressure is still necessary to trigger action in many instances. The studies also reported a general lack of formal collaboration between sectors, and lack of systematic use of indicators. Major impediments to the decision-making process include lack of resources, lack of political motivation, lack of technical capacity, lack of a participatory approach, and lack of or inadequate use of data.

These findings have considerable significance for local intervention and management of environmental health problems. On the one hand, it is crucial that the relevant authorities have the obligation, capability and information necessary to act to protect human health, and as such should do so pre-emptively. On the other hand they should not be immune to the concerns of the local community. Effective decision-making requires an efficient two-way process of communication between the authorities and the community. The community should be informed about the environmental health priorities and actions of the authorities, and in turn should be able to influence those priorities and actions by voicing its own concerns. Indeed, one of the main purposes of the HEADLAMP process is to foster this sort of collaboration.

Use of indicators for decision-making
The use, or lack of use, of environmental health indicators was identified and discussed in all the study sites. The use of indicators is relatively well-established in the Philippines, for example in the legislation to reduce lead in petrol. Indicators are also used in Chile — the Basic Law of the Environment implies the use of indicators to monitor compliance with emissions and environmental quality standards, while the Environmental Recovery Plan for

Talcahuano is monitored using indicators. In several of the cities, however, indicators were not employed in any systematic way. Use of indicators largely reflected the relative availability of environmental and health data, and the maturity of the local decision-making systems within these cities. A "chicken-and-egg" situation can thus be envisaged where indicators offer a means of encouraging and targeting data gathering and improving local decision-making, but it is likely that the use of indicators often occurs only as a result of such advances.

Part of the field studies consisted of testing a set of 17 environmental health indicators (Box 8.2). These indicators had previously been proposed as a basis for monitoring the implementation of Chapter 6 of Agenda 21, which deals with health in the context of sustainable development (a modified set was eventually adopted by the United Nations; see United Nations, 1996). The purpose of including them in the field studies was to assess whether indicators proposed for use at the national level were of relevance and available at the local (i.e. city) level. The studies found that the following indicators were either available or could feasibly be collected at the local level in all the study areas (the Managua study did not include an evaluation of these indicators and is therefore not included in these findings):

- Infant mortality rate (per 1,000 live births).
- Maternal mortality rate (per 1,000 live births).
- Percentage of the population with access to sufficient quantity of safe drinking water.
- Percentage of newborn weighing at least 2,500 g at birth.
- Percentage population covered by primary health care.
- Percentage of the eligible population that have been fully immunised according to national immunisation policies.
- Adult literacy rate.

Prevalence of malaria was collected where relevant. The indicator "Percentage of the population with access to hygienic excreta disposal facilities" was not collected (and was found to present serious difficulties) in Calcutta only. The following indicators were problematic (not collected and/or hard to collect) in most settings:

- Percentage of people served by public waste removal service.
- Percentage of people exposed to high concentrations of health damaging air pollution (outdoors and indoors separately).
- Percentage of people without access to adequate food supply.
- Prevalence of intestinal helminths among children (aged 2 to 15).
- Incidence of acute enteric infections.
- Life expectancy at birth.

The indicators relating to health expenditures (percentage gross national product spent on health; and percentage of national health expenditure devoted to primary health care, health centres and regional hospitals) were not found to be in use locally.

The DPSEEA framework would appear to be a useful and rationalising tool in relation to local environmental health issues. It was clear in the case study areas that the available information and indicators often tended to focus mainly on the "downstream" parts of this framework, that is in relation to the state of the environment, exposures, effects and actions. If decisions are to be taken from a longer-term perspective and in a more integrated way, there is a need to set issues within a wider context. Information on the driving forces and pressures would help to achieve this. Use of this framework was encouraged in the field studies. In the Philippines, it was subsequently proposed by the Department of Health as a potential basis for developing a risk management plan for environmental health impact assessment (Department of Health, Philippines, 1997).

Stakeholder involvement in decision-making
The need to identify and involve all relevant parties (and sectors) as part of the HEADLAMP process has already been emphasised. This is important not only to help identify and prioritise the environmental health issues of concern, but also to facilitate access to relevant data, and to bring together those who hold the capability to resolve the problems, either on their own or collectively. Often, the health sector does not have the authority or expertise to address problems at their source — such action lies within the responsibility of other sectors both of government (e.g. environment, industry or transport departments) and business. The public also have considerable scope to reduce environmental health risks, for example through their changes in their own risk behaviour (e.g. changes in consumption patterns) and through the pressures they can exert on other decision-makers.

Identifying and bringing together this range of stakeholders is nevertheless challenging. The range of relevant stakeholders may vary greatly from one environmental health issue to another. Many vital stakeholders (e.g. young children, the poor, the infirm) may also not be able easily to represent themselves. Whether they can be represented, thus depends on whether appropriate pressure groups or community groups exist. One of the most important interest groups, future generations, clearly cannot speak for itself and needs to be considered by proxy, by all those involved.

Faced with these challenges, the field studies generally identified the main professional sectors and parties that have a stake in each of the environmental health problems identified, even though these were not necessarily canvassed or consulted directly. The identification of the key community groups which need to be involved as part of this process was, however, considerably less successful. Only three of the field studies conducted workshops, and none of these managed to bring together all the relevant parties, and in each case specific problems were encountered in involving local communities. In Talcahuano, community attitudes to participation (a product of limited local empowerment) undoubtedly contributed to the difficulty in involving a wider range of stakeholders. In Cape Town there is awareness of the need to integrate previously excluded groups into the decision-making process, although this clearly poses difficulties because of their lack of experience. Managua is politically divided, making community participation difficult. Such difficulties are inherent in any attempt to promote local participation, and clearly need to be addressed if health and environment problems are to be tackled at the local level.

In spite of these problems, the workshops which were run as part of these studies proved to be useful. Many of the studies identified a lack of co-ordination and co-operation between the different sectors. This is one of the major impediments for environmental health management. The workshops were an important mechanism to bring these disparate sectors together and to help foster a shared understanding of the issues involved.

8.6   Conclusions

The field studies outlined in this chapter clearly demonstrate the potential of the HEADLAMP approach to fostering and supporting local environmental health decision-making. In each case, the studies helped to identify local environmental health problems, to define the availability of and needs for information, and to begin the process of bringing relevant stakeholders together in a collaborative spirit. To be wholly effective, however, the approach needs to be more deeply embedded within the local decision-making system. This will depend upon two main preconditions: the existence of clear political will to tackle environmental health issues at a local level; and specific action to strengthen the availability and relevance of local information on environmental health.

The development of political will is a key prerequisite. It is likely to be encouraged in part by sharing the experience and seeing the benefits from actions such as HEADLAMP projects. It is also likely to develop in response to wider shifts in national health planning towards a more decentralised, local level system, as has happened in the Philippines. Moreover, once such a principle has been established, it is likely to be largely self-sustaining, because it will help to establish both an expectation of local involvement in decision-making, and the existence of groups which can represent the various stakeholders effectively when the need occurs.

The availability of relevant data on environment and health is equally important. Until recently, attempts to bring these data together have largely relied on the efforts of researchers in academic institutions (and then primarily for exploratory reasons and on the basis of special surveys). Routine data are often not considered appropriate for research purposes and thus they are either ignored or supplemented with additional data collection. There is, however, considerable value in linking routine environmental and health data at the local level specifically to raise local awareness about environmental health issues and to support decision-making. For the long term, efforts to improve routine data gathering are needed to be given immediate priority for this purpose. Where relevant data do not exist, rapid survey methods need to be applied. In many areas, this will require significant capacity-building to help design appropriate monitoring and information gathering systems, to help construct the relevant infrastructure (including monitoring, data analysis, mapping and reporting systems) and to help train personnel. These and other requirements involved in full implementation of the HEADLAMP process are discussed in the next chapter.

8.7   References

Briggs, D., Corvalán, C. and Nurminen, M. [Eds] 1996 *Linkage Methods for Environment and Health Analysis - General Guidelines*. World Health Organization, Geneva.

Briggs, D.J., Stern, R. and Tinker, T.L. [Eds] 1998 *Environmental Health for All. Risk Assessment and Risk Communication for National Environmental Health Action Plans.* NATO Science Series. 2. Environmental Security - Vol. 49, Kluwer, Dordrecht.

Corvalan, C. and Kjellstrom, T. 1997 Analysis of health and environment indicators for decision-making: Lessons from field studies. *Urbanisation and Health Newsletter*, No. **33**.

Corvalán, C., Nurminen, M. and Pastides, H. [Eds] 1997 *Linkage Methods for Environment and Health Analysis - Technical Guidelines*. World Health Organization, Geneva.

Department of Health, Philippines 1997 *Philippine National Framework and Guidelines for Environmental Health Impact Assessment*. Department of Health, Manila.

Fink, A., Kosecoff, J., Chassin, M. and Brook, B.H. 1984 Consensus methods: characteristics and guidelines for use. *American Journal of Public Health*, **74**, 979–83.

Gonzalez, M., Barten, F. and Sanchez, A. 1995 Health and Environment Analysis for Decision-making (HEADLAMP) field study: Managua. Unpublished document, World Health Organization, Geneva.

Instituto Nacional de Estadísticas 1977, 1985, 1991, 1992 *Compendio Estadístico.* Instituto Nacional de Estadísticas, Santiago.

Instituto Nacional de Estadísticas 1995 *Anuario de Demografía.* Instituto Nacional de Estadísticas, Santiago.

Jones, J. and Hunter, D. 1995 Consensus methods for medical and health services research. *British Medical Journal*, **311**, 376–80.

Kjellström, T. and Rosenstock, L. 1990 The role of environmental and occupational hazards in the adult health transition. *World Health Statistics Quarterly*, **43**, 188–96.

Lewin, S. 1995 Improving decision-making for environmental health in Cape Town - the HEADLAMP field study: Summary of interim findings and future directions. *Urbanization and Health Newsletter*, No. **26**.

Lewin, S. 1996 Health and Environment Analysis for Decision-making (HEADLAMP) field study: Cape Town. Unpublished document, World Health Organization, Geneva.

Mukherjee, A., Chakrabarti, C., Majumdar, A. and Chatterjee, J. 1995 Health and Environment Analysis for Decision-making (HEADLAMP) field study: Calcutta. Unpublished document, World Health Organization, Geneva.

Murray, C.J.L. and Lopez, A.D. [Eds] 1994 *Global Comparative Assessments in the Health Sector*. World Health Organization, Geneva.

Murray, C.J.L. and Lopez, A.D. [Eds] 1996 *The Global Burden of Disease: a Comprehensive Assessment of Mortality and Disability from Diseases, Injuries and Risk Factors in 1990 and Projected to 2020*. Harvard University Press, Cambridge, Massachussets.

Richey, J.S., Mar, B.W. and Horner, R.R. 1985a The Delphi technique in environmental assessment - I. Implementation and effectiveness. *Journal of Environmental Management*, **21**, 135–46.

Richey, J.S., Horner, R.R. and Mar, B.W. 1985b The Delphi technique in environmental assessment - II. Consensus on critical issues in environmental monitoring program design. *Journal of Environmental Management*, **21**, 147–59.

Salinas, M., Vega, J., San Martin, A. and Manriquez, S. 1995 Health and Environment Analysis for Decision-making (HEADLAMP) field study: Talcahuano. Unpublished document, World Health Organization, Geneva.

Smith, K. 1990 The risk transition. *International Environmental Affairs*, **2**, 227–51.

Smith, K. 1997 Development, health, and the environmental risk transition. In: G.S. Shahi, B.S. Levy, A. Binger, T. Kjellstrom and R. Lawrence [Eds] *International Perspectives on Environment, Development and Health: Toward a Sustainable World*. Springer Publishing Co., New York, 51–62.

Smith, K., Corvalán, C. and Kjellström, T. 1999 How much global ill health is attributable to environmental factors? *Epidemiology* (in press).

Songsore, J. and Goldstein, G. 1995 Health and environment analysis for decision making (HEADLAMP): field study in Accra, Ghana. *World Health Statistics Quarterly*, **48**, 108–117.

Soton, A., Alihonou, E., Gandaho, T. and Defonsou, M. 1995 Health and Environment Analysis for Decision-making (HEADLAMP) field study: Cotonou. Unpublished document, World Health Organization, Geneva.

Soton, A., Alihonou, E., Gandaho, T. and Defonsou, M. 1997 Environmental health indicators for decision-making: a case study in Cotonou, Benin. *Urbanisation and Health Newsletter*, No. **33**.

Stephens, C., Akerman, M. and Borlima, P. 1995 Health and Environment in Sao Paulo, Brazil: methods of data linkage and questions of policy. *World Health Statistics Quarterly,* **48**, 95–107.

Torres, E. and Subida, R. 1996 Health and Environment Analysis for Decision-making (HEADLAMP) field study: Manila. Unpublished document, World Health Organization, Geneva.

UNDP (United Nations Development Programme) 1990 *Human Development Report 1990.* Oxford University Press, New York/Oxford.

UNDP (United Nations Development Programme) 1997 *Human Development Report 1997.* Oxford University Press, New York/Oxford.

United Nations 1993 *Agenda 21: Programme of Action for Sustainable Development.* United Nations, New York.

United Nations 1996 *Indicators for Sustainable Development Framework and Methodologies.* United Nations, New York.

Victorin, K., Hogstedt, C., Kyrklund, T. and Eriksson, M. 1997 Setting priorities for environmental health risks in Sweden. In: *Proceedings from the annual meeting of the Society for Risk Analysis - Europe.* Stockholm School of Economics, Center for Risk Research, Stockholm, 211–22.

Victorin, K., Hogstedt, C., Kyrklund, T., Eriksson, M. 1998 Setting priorities for environmental health risks in Sweden. In: D.J. Briggs, R. Stern and T.L. Tinker [Eds] *Environmental Health for All. Risk Assessment and Risk Communication for National Environmental Health Action Plans.* NATO Science Series. 2. Environmental Security - Vol. 49, Kluwer, Dordrecht, 35–51.

WHO 1995 HEADLAMP Field studies protocol. Document No. WHO/EHG/95.29, World Health Organization, Geneva.

WHO 1997 *Health and Environment in Sustainable Development: Five Years After the Earth Summit.* World Health Organization, Geneva.

World Bank 1993 *World Development Report 1993: Investing in Health.* Oxford University Press, New York and Oxford.

Chapter 9[*]

THE HEADLAMP APPROACH: A NEW MODEL FOR ENVIRONMENTAL HEALTH DECISION-MAKING

9.1   Information, decision-making and action

In a world in which the population is fast becoming increasingly urbanised, in which technological and economic development is happening apace, and in which the balance between environment and health is coming under increasing strain, there is a growing need for new approaches to environmental health decision-making which can help to protect and improve the health of people in all areas of the world. The HEADLAMP methodology is aimed at providing such an approach. It represents an attempt to develop and apply a new model of environmental health decision-making which can improve public health not just as a one-off initiative, but by establishing long-term partnership between those involved, and by providing a firm information base for debate, management and policy.

Several principles and assumptions underlie the HEADLAMP approach. First and foremost, those concerned must genuinely use information to guide and support their decisions — information must lead to action. Second, this information must be relevant, balanced and reliable; it must go beyond partiality and opinion and provide sound and defensible evidence for action. Third, the approach must be holistic — it needs to set decisions within the wider context of causes and effects, so that the actions taken can be co-ordinated and integrated effectively and problems can be dealt with collectively, rather than as a set of separate and very specific issues. Fourth, it must be proactive and preventative: it must help to detect problems before they become acute and it must help to take action which avoids, rather than merely ameliorates, adverse health effects and promotes positive health outcomes. Finally, it must be inclusive, in that it should actively and fairly involve all the stakeholders concerned in ways which help to build consensus about the actions that are needed.

Each of these principles and assumptions has many implications, and raises both conceptual and practical questions. In this final chapter, therefore,

---

[*]   *This chapter was prepared by D. Briggs, G. Zielhuis and C. Corvalán*

some of these issues are considered in order to expand the philosophy under-pinning the HEADLAMP methodology and also to examine the conditions which need to be in place in order to ensure its success.

9.2   Evidence begets action

The HEADLAMP approach is aimed at strengthening the information basis for local decision-making. It is, essentially, a way of undertaking local research, and collating local knowledge, in order to make information on environmental health issues available to decision-makers. It derives from the assumption that a better understanding leads to better decisions and that information does, indeed, beget action.

This assumption is fundamental to the HEADLAMP approach, and may seem self-evident. In practice, however, it cannot be taken for granted. Whether decision-makers use information to guide their decisions and, if so, how and at what stage in the process, is the subject of fierce debate. The "rational" model of decision-making is now under challenge, partly because in the light of post-modernist thinking it has been recognised that the rela-tionships between information, knowledge, decisions and actions are far more fluid and far more value-laden than has traditionally been assumed (see for example Rivett, 1994).

Certainly, most decisions are taken on the basis of information: otherwise the process would be no more than a blind and random activity. The quality and comprehensiveness of this information may nevertheless leave much to be desired. Decision-makers do not usually have complete information. Much is often missing because the relevant monitoring or information gath-ering has not been undertaken. They may not even have all the information that is available, either because they are not aware of, or cannot gain access to, much of what does exist. In addition, they may not use all the information they do have: much of it may be rejected because it is difficult to understand, does not seem well-founded, or perhaps does not accord with preconceived views. Simple overload of information often means that a lot of it is ignored. Access to and use of information thus vary greatly, depending not only on the level of information available (i.e. the quality of the data-gathering systems) but also the political culture in which the decisions are made and the personal characteristics, expertise, experience and attitudes of the decision-makers. Above all, information is never value free; the existence of information reflects a previous, value judgement about the merits of collecting it. The selection of what information to use derives from the values of the user. The use of that information involves interpretations, which are themselves set within a framework of cultural and personal values. As Rivett (1994, p. 261) thus states "*The decisions taken by management tell one a great deal about the latent value system, and the data which management uses tell one much about the management itself'.*

As this also implies, not all information drives action; often the reverse is true. It has already been noted, for example, that information is costly and information is not always easy to obtain. Investment in information gathering thus often occurs only when there is a clear justification and need. The decision to make this investment is, therefore, to a large extent a managerial or political one. It is likely to be made only when a political need already exists: in other words when a problem or the potential for a problem has already been acknowledged to exist. In these cases, information may be gathered not to help make decisions and guide action, but to justify decisions that have already been made and to assess their effects. Nonetheless, the knowledge that derives from this information may ultimately contribute to future decisions and help to stimulate new actions.

The relationship between information and action, therefore, is perhaps best seen not as a linear one, flowing either in one direction or the other, but as a much more complex and reiterative decision system (Figure 9.1). Information, much of it circumstantial and experiential, helps to create a picture of the existence of a problem and the need for action. An initial assessment of this issue may then be made. If it is seen as serious, action may be taken without any further information gathering. If doubts exist, or if for political or other reasons delay is seen to be expedient, further information gathering may be initiated. The decision to act, itself, may also generate the demand for more information, for example to help rationalise the decision and to monitor the effects of intervention. This process adds to the new body of partial and circumstantial evidence and as such contributes to the emergence of new questions and concerns, and new cycles of decision-making. Throughout this process, therefore, multiple decisions may be taken, and repeated information inputs may be necessary.

This more complex model of decision-making does not undermine the value of the HEADLAMP process — if anything, it emphasises its role. The HEADLAMP approach gives a means of adding new information at different points within the cycle. It provides background, contextual information that can help to identify possible issues and pose new questions. It provides indicators and methods to help assess and prioritise these issues more rationally at the decision-making stage. It also offers a means of monitoring environmental and health conditions following intervention, as a basis for assessing the effects of the actions taken, and to help steer the actions more effectively. In addition, it aims to ensure that all this information is drawn widely from all the stakeholders involved, as well as all the sources of relevance, in order to encourage more balanced and less partial decisions.

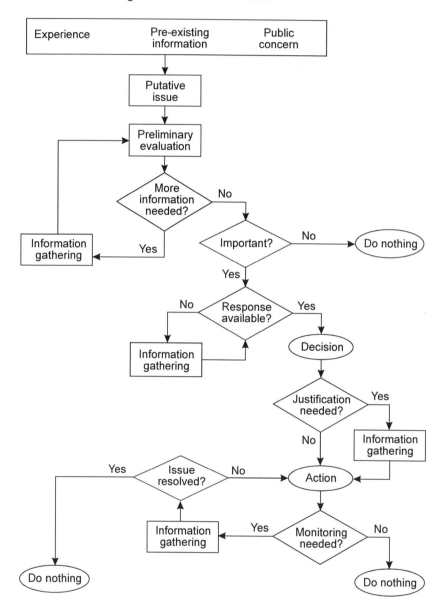

Figure 9.1 A model of the relationship between information and decision-making

It is apparent from the discussion above that the effectiveness of the HEADLAMP process depends fundamentally on the willingness and capability of those concerned to use the information available, and to be guided by what it shows. Information achieves little if it is merely stored and ignored. Important determinants of action are thus the confidence, autonomy and skills-level of managers and other stakeholders. Where these are not adequate, the first need may not be to improve the availability of information, but the training and empowerment of those concerned.

9.3   A sound basis of evidence

The very need for the HEADLAMP approach derives from the circumstance that the information needed to guide decisions does not already exist, or cannot readily be obtained. One of the main purposes of HEADLAMP projects is to fill this information gap. If actions are to be guided by this information, however, then it must provide a sound basis of evidence.

As noted in Chapter 5, three types of information are required: environmental information, health information and information on the relationship between the two. Each of these poses specific challenges and problems. Environmental information, for example, should include data on the exposures that are likely to affect health and, because action is often targeted at source, on the causes and sources of these hazards. Difficulties arise in this case because of the wide range of hazards for which information might be needed (e.g. the many different pollutant species), the different environmental media and exposure pathways involved, and the large degree of temporal and spatial variability that occurs in the environment. Against this background, environmental monitoring is remarkably sparse, even in more developed countries. Direct measures of exposure are rarely available, such that exposure often has to be deduced from other information — for example on environmental concentrations or levels of source activity. Data on environmental conditions, however, are also limited, and often restricted to a small number of (not necessarily) representative monitoring or survey sites. Data on source activity (e.g. traffic volumes, employment in industry) may be more readily available, but as measures of exposure they are far from ideal because of their remoteness from the actual exposure event. As noted in Chapter 5, therefore, exposure assessment is often approximate at best, and may not be adequate to define accurately the population at risk.

Data on health are often similarly variable in both availability and quality. Crude mortality data are available in many cases, and routine data may also be available for some types of morbidity (e.g. some communicable diseases and cancers). Problems nevertheless exist in relation to the accuracy and consistency of diagnosis, coding, spatial referencing and level of aggregation of these data. Information on most forms of morbidity are generally lacking or substantially incomplete.

Information of the association between environmental exposures and health (i.e. the dose–response or exposure–effect relationships) is often no less uncertain. This information is needed in order to quantify the health risks associated with any environmental hazard: for example to provide a measure of the number of people at-risk, or the number of additional cases of morbidity or mortality as a result of exposure. Unfortunately, very few well-established dose–response relationships have yet been derived for environmental risk factors. For the vast majority of the hazards currently of concern, considerable uncertainty still prevails. In part, this derives from the small relative risks involved. Taubes (1995, p. 164) quotes Michael Thun, the director of analytic epidemiology for the American Cancer Society: "*With epidemiology you can tell a little thing from a big thing. What's very hard is to tell a little thing from nothing at all*". Yet low relative risk characterises many of the problems in environmental health, especially those associated with chronic exposures. Many of the relative risks reported by recent studies of respiratory diseases in children, for example, are less than 2 (Table 9.1). This places these studies well within the area of analytical uncertainty. Indeed, in the case of childhood asthma, there have been many recent studies that have shown no effect of exposures to traffic-related pollutants. Nevertheless, low relative risk does not imply a negligible effect. As Taubes (1995, p. 164) states: "*...these subtle risks — say the 30% increase in the risk of cancer from alcohol consumption that some studies suggest — may affect such a large segment of the population that they have potentially huge impacts on public health*". Thus, while the relative risk may be low, the impact across the whole population may be large.

Against this background, consensus on the risks associated with exposures to environmental hazards is generally lacking, and few widely accepted dose–response relationships exist. In applying the HEADLAMP process in many areas, it will often be necessary, therefore, to derive dose–response relationships from the available literature. Meta-analysis is a potentially powerful tool in this respect. Nevertheless, there are a number of problems with this approach. One problem is publication bias: there is a definite tendency for positive studies to be reported more readily than negative studies, even though the latter are not necessarily based on less rigorous analysis. Thus published papers cannot be regarded as a random sample of the relevant research. Another problem is that there is often little consistency in different studies: they may use different study designs, different exposure indicators and different health endpoints. This makes comparisons of different studies difficult.

In the absence of agreed dose–response relationships, an alternative is to use established exposure limits or environmental thresholds to identify those at risk. The WHO environmental quality guidelines for air pollutants (WHO, 1987), water pollutants (WHO, 1993), food (FAO/WHO, 1989) and the workplace (e.g. WHO, 1980, 1986) are useful in this respect. National environmental standards may also be available. These are, of course, based upon epidemiological and other studies which have examined the link between

Table 9.1 Risk factors for respiratory diseases in children

Location	Study design	Date	Exposure indicator	Health outcome	Odds ratio	Reference
3 areas, Italy	Cross-sectional	1992	Outdoor air pollution	Asthma	1.3–1.4	Forastiere et al., 1992
Ashod & Hadera, Israel	Cross-sectional	1988	High/low air pollution	Cough without cold Asthma	1.5 2.7	Goren and Hellerman, 1988
Haifa Bay, Israel	Cross-sectional	1984	High/low air pollution	Sputum with cold Sputum without cold	1.4 1.8	Goren et al., 1990
Northern Finland	Follow-up	1982	SO_2, particulates, NO_2	Respiratory infection	1.6–2.0	Jaakkola et al., 1991
	Ecological	1986–88	PM_{10} SO_2	Respiratory mortality Respiratory mortality	2.4 3.9	Bobak and Leon, 1992
Sheffield, UK	Case-control	1991	NO_2	Wheeze	1.0 0.8	de Hoogh, 1999
London, UK	Case-control	1992–94	NO_2	Hospital admissions for respiratory illness	1.0	Wills, 1998
Huddersfield, UK	Ecological	1994	NO_2	Cough in last 12 months	0.9	Elliott et al., 1995
Amsterdam, Netherlands					0.8	
Prague, Czech Republic					0.9	

exposure and health effect. Unlike formal dose–response relationships or odds ratios, however, they also invariably involve some degree of political judgement (e.g. based on the costs and political acceptability of implementation). In either case, this approach has some limitations. In particular, it fails to reflect the increased risks that may occur where these guidelines are exceeded (i.e. the standards do not reflect the full dose–response relationship).

Whichever approach is taken, care is also needed in translating either dose–response relationships or environmental guidelines to new areas, because they have often been established in specific social, environmental and health care situations. Whether they are valid elsewhere depends on whether the conditions likely to affect the dose–response relationship are comparable. Often, this is not the case and substantial differences may exist, for example, in the background health status of the population, the exposure range, the mix of hazards to which people are exposed, and the quality of and access to health treatment. One advantage of collating background information on the area as part of the HEADLAMP process is that it allows these comparisons to be made. Wherever possible, local validation of dose–response relationships should also be undertaken, if not through formal epidemiological studies then by collecting and comparing data on both environmental exposures and health effect. This is another reason why the HEADLAMP approach is based on the concept of data linkage.

Acquiring information which provides a sound basis of evidence is therefore often problematic, especially in developing countries where the established infrastructure for monitoring and research is limited. Again, this does not negate the value of the HEADLAMP approach, but rather the reverse. In these circumstances, one of the main functions of the HEADLAMP process is to help identify the gaps and uncertainties in the available information, and to encourage new data collection where it is required. Some of the rapid survey methods developed by WHO (e.g. WHO, 1982; Economopolous, 1993) are especially valuable for this purpose. At the same time, it is important to recognise issues of data quality, where data do exist. Good practice in applying the HEADLAMP approach thus involves making explicit statements about uncertainties in the data and in any derived estimates of risk, so that decision-makers can make valid judgements about the reliability of the information. In this context it nevertheless needs to be acknowledged that people's understanding of risk and uncertainty is often poor, whether they are professionals or lay people. Effective development of the HEADLAMP approach may involve training and education of those concerned about concepts or risk, uncertainty and the implications for interpreting information.

9.4   A holistic approach

The links between environment and health are evidently complex. Individually, people are exposed to a wide range of environmental pollutants and other risk factors, at different times and places, and each person has different susceptibilities to their effects, and differs in terms of their access to health treatment and care. Moreover, the various risks to which people are exposed do not operate wholly independently, but interact and have synergistic effects. Specific pollutants may derive from many different sources; individual sources produce a wide range of pollutants; any pollutant may move through the environment in a myriad of different ways; and the effects of exposure may vary depending on the characteristics of the people exposed. The environment–health relationship is thus characterised by multiple causes, multiple pathways and multiple effects. For these reasons, a holistic approach is needed for identifying priorities for action and for designing the actions that need to be taken. Discussing the issue of pollution, for example, Dunn and Kingham (1996, p. 838) argue "*...relationships need explaining not only in terms of specific pollutants but also in relation to interactions between different pollutants, both from single sources and from multiple sources: the 'cocktail' effect. The question, then, is: how do we tease out these multiple and interactive effects operating at the individual and community level, and beyond, in order to draw useful conclusions to explain health status in populations exposed to a variety of environmental and socio-economic risks?*"

Unfortunately, the epidemiological and other sciences that often form the basis for decision-making are themselves far from holistic in approach, but tend to be strictly reductionist and based on a "single cause–single effect" perspective. As a result, knowledge of the collective effects of different exposures on health is often limited. Interaction may lead to non-additive effects, which cannot be assessed simply by accumulating risks from different hazards. The multiplicity of sources, pathways and agents means that, in investigating any single agent or exposure, it is likely that only a small part of the overall problem is being seen. This means that it can be dangerous simply to extrapolate from existing information.

This problem of reductionism is, paradoxically, less of a constraint in terms of management and policy actions. Many of these interventions are far blunter and more aggregate tools, and even when actions are directed at specific concerns or risks, they may have far-reaching consequences. Traffic management schemes, for example, might be introduced in response to concerns about road accidents; at the same time, however, they may have a number of other health benefits including reducing exposures to fine particulates, nitrogen dioxides, benzene, and a range of other air pollutants, as well as noise and vibration — whilst also helping to reduce congestion and community severance. Similarly, action to improve access to safe water in the home may help not only to reduce exposures to waterborne diseases, but also may contribute to improved domestic hygiene and sanitation and reduced problems of food-borne disease, as well as saving large amounts of time and labour in water-carrying (this time and labour can then be put to other uses).

As such, policy actions have important advantages in relation to environmental health risks, and important disadvantages. On the positive side, they perhaps reflect more realistically the complex interactions and interdependencies that occur in the real world. Rather than tackling risk factors individually, therefore, they often address them collectively. In the process, they resolve not one problem at a time, but several problems simultaneously. On the negative side, however, it is clearly difficult to predict or evaluate the effects of such non-specific policies, largely because little information is available on the relationships with health outcome at this aggregate level. This is one of the main reasons why, in the HEADLAMP approach, indicators are sought throughout the DPSEEA chain, including on the wider socio-economic and policy context. It is also why linkages between indicators are stressed. In this way, the indicators provide a means of assessing and monitoring the interdependencies between action and health outcome, and of evaluating the combined effects of interventions on the wider realm of health.

9.5   A preventative approach

Policy action can be taken at many different points in the environmental health chain, as the DPSEEA framework (Figure 3.4) shows (see also Figure 9.2). In environmental health, however, as in other areas of policy, prevention is usually better than cure. This is true not only because prevention can avoid unnecessary human suffering, but also because it is often more cost-effective to implement (it provides major savings in health service costs and lost work-time and productivity). If the HEADLAMP approach is to be effective, therefore, it should contribute not just to the identification of health effects retrospectively, but equally to the prediction and prevention of future risks.

In order to achieve this, HEADLAMP needs to be able to provide clear information on impending risks. One way of doing this is to extrapolate effect-based indicators to predict future levels of morbidity or mortality. For example, the rising trend in childhood asthma, seen in most countries across the world, may be taken to imply that this is a problem requiring action. This approach nevertheless has many weaknesses. The first is that trends are often difficult to detect, due to uncertainties in the data or the complexities of temporal variations in the incidence of the health effect (e.g. due to the influence of short-term seasonal or episodic variations). Secondly, this approach can only work effectively where the data are available in a suitably timely

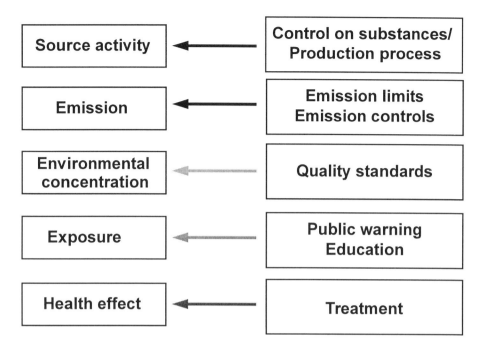

Figure 9.2 Points of intervention in the environment–health chain. Note the strength of the arrow represents the general effectiveness of the intervention (After Briggs, 1998)

fashion: i.e. where the data are updated and made available on a timescale considerably shorter than the trends of interest. Otherwise, by the time the trend is apparent, the damage will already be done. The use of health trend data is also less effective with diseases characterised by long latency periods (e.g. many cancers), because the warning of a problem will only occur too late for preventative action to be taken. In addition, analysis of trend data in health outcome does not, by itself, indicate where action is needed. This requires an understanding of the causal links and exposure pathways involved.

A more effective approach, in many cases, is to use exposure-based indicators (e.g. on exposures, state of the environment, pressures or driving forces) to provide an early warning of possible health effects. This is doubly useful because it also helps to target attention at the source of the problem, rather than the effect. Again, however, this is only possible if several preconditions exist. One need is for reliable and up-to-date information on the exposures or source activities of interest. Once more, this emphasises the importance of the timeliness of the data. A second requirement is for a clear understanding of the relationship between these risk factors and health. Ideally, this association should be expressed as a dose-response relationship, because reasonably accurate predictions of health outcomes can then be made, for different exposure scenarios. Nevertheless, even a more qualitative understanding of the relationship can help to indicate the general magnitude of the potential problem, and thus help to assess and prioritise the need for action. Unfortunately, as noted previously, knowledge of the relationships involved is often weak — especially in the case of very remote and non-specific sources or activities.

In either case, it also needs to be recognised that prevention cannot be seen as a "one-off" process. If effective action is really to be taken to prevent adverse health effects (and, by corollary, promote positive health outcomes), then it needs to be informed by a continuous process of surveillance and review. The HEADLAMP process should not be regarded, therefore, as a singular and isolated event — a quick snapshot of the environmental health situation followed by a once-and-for-all set of decisions. Instead, it is intended to be a continuous and self-sustaining process. It should help to establish, and be embedded within, a regular and routine process of information gathering, assessment and review.

9.6   An inclusive approach

Concern and responsibility for environmental health are not just the prerogative of policy-makers and officials. They belong to everybody, if not as decision-makers in any real sense, then as victims or taxpayers. If decisions on environmental health are thus to be inclusive (if they are going to take account of the interests of all the stakeholders, and actively involve all those with some power to influence the outcome) then they require consultation and collaboration with a wide range of people and institutions. Indeed, both the National Academy of Sciences (1991) and the Presidential/Congressional Commission on Risk Assessment and Risk Management (1997) have recently emphasised that individuals have an entitlement to be consulted in matters relating to public health. One of the fundamental objectives of the HEADLAMP approach is to encourage and facilitate this inclusive approach.

Involving the many different stakeholders with interests in environmental health has major benefits. It helps to make sure that actions can be implemented more effectively — by those who are most appropriate, and at an appropriate point in the environment–health chain. It should also help to ensure that decisions are taken on the basis of consensus rather than from a position of partiality and power, and that the actions are approved and supported by all those concerned. Without access to shared information, these various stakeholders are likely to be influenced largely by individual experience or prejudice. Even with information, however, consensus may be hard to achieve. Information is not unitary: it can be interpreted and used in many different ways. Consensus building thus requires not just the one-way communication of knowledge from those who believe they know (usually the

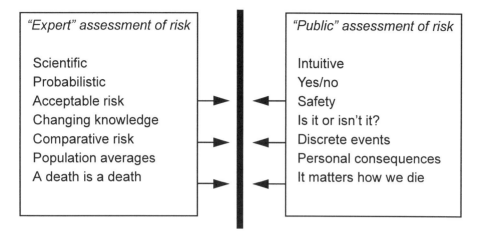

Barrier to mutual understanding

Figure 9.3 The different languages of risk (After Powell and Leiss, 1997)

professional) to those who "need to be told" (the public). Rather, it requires a two-way exchange of information, and the opportunity to debate and negotiate, to challenge and respond. Effective, local decision-making is thus an interactive process between the many stakeholders concerned.

Problems nevertheless occur in trying to bring together people from different backgrounds, and with different interests, in this way. One of the main problems relates to the different levels of understanding, and the different belief systems, of those involved (Farago, 1998). Major differences may exist, for example, in the very concepts and language of risk between professionals and the public (Jardine and Hrudey, 1998a); Powell and Leiss (1997) draw the distinction between the language of the "experts" and the language of the "public". The former based on specialised knowledge, usually reliant on published scientific literature and couched in technical terminology; the latter intuitively grounded, experiential and using colloquial terms (Figure 9.3). Put together, these two conceptual systems can lead to mistrust and frustration. The experts fail to see why the public cannot understand the complexities and uncertainties in their data, and the statistical basis of their risk estimates. The public expect the science to be unambiguous, unequivocal and clear. At the same time they expect experts to understand that they are concerned not just about the actuality of harm, but the fear of harm, not just to themselves but to their loved ones as well; and not just because they necessarily expect to be harmed, but because they would feel responsible for any harm.

Attitudes to risk are thus, to a large extent, culturally determined and experience-based. Douglas and Wildawsky (1982), for example, suggested that those who support risk-taking in any situation tend to occupy a central position within their society, while those who oppose risk have a marginal position in the social structure. Pages *et al.* (1988) found that personal factors, such as qualifications, age, political attitudes, occupation and social status, were related to people's social values and that these, in turn, influenced attitudes towards risks.

All this implies that the methods of risk communication should be designed and segregated according to the needs of those involved. If this process of communication is to comprise a real dialogue, on equal terms, between all those involved, it poses major challenges, because it clearly requires that the barriers to mutual understanding are broken down and that trust is established between the various stakeholders involved. This means that the relevant people must be involved, in an appropriate way, and at the right stage in the HEADLAMP process. Sadly, these requirements are often not met and, as Jardine and Hrudey (1998b) point out, efforts at public participation often fail because the interested and affected parties are brought into the process after the problem has been defined and characterised, or because they are asked to participate in a decision-making process that does not reflect their own concerns and fears.

The U.S. National Research Council (1996) stresses five objectives for effective public participation:

- *Getting the science right.* High scientific standards must be attained in relation to the measurements, analytical methods, data sources, assumptions and acknowledgement of uncertainty.
- *Getting the right science.* The significant risk-related concerns of all parties should be addressed.
- *Getting the right participation.* Participation must be broad enough to ensure that important, decision-relevant information enters the process, that all important perspectives are considered and that the legitimate concerns about inclusiveness and openness of those concerned are met.
- *Getting the participation right.* The process must be able to convince those concerned that it is responsive to their needs; that their information, views and concerns have been adequately represented and recognised; that they have been adequately consulted; and that their participation has been able to affect the way risk problems are identified and understood.
- *Developing an accurate, balanced and informative synthesis.* The process must reflect the state and range of relevant knowledge, and satisfy all parties involved in the decision that they have been adequately informed within the limits of available knowledge.

Four crucial parts to the participation process may also be defined: identification of the relevant participants; collation of knowledge and information from those concerned; debate amongst those concerned about the relative importance of the various issues and the actions that need to be taken; and the supply of information, including feedback on all decisions and outcomes, to those involved.

A wide range of methods are available which can contribute to one or more parts of this process, some of which were applied in the HEADLAMP case studies reviewed in Chapter 8. These include local workshops, focus groups, citizens panels, phone-in events, Internet sites and electronic conferences, questionnaire and interview surveys, public presentations and newsletters. An important role in many of these activities is played by the media, which provides the main source of information on environmental health issues for most people, although it is often seen as being neither balanced not well-informed in its treatment of these issues and is not considered especially trustworthy (Jardine *et al.*, 1995). Special efforts may therefore be needed to involve the media early in the HEADLAMP process, and to raise their understanding of the issues involved by detailed briefings and practical demonstrations.

Debate and negotiation about the issues raised may also need the use of additional tools and methods. Effective presentation and communication of scientific information is an important requirement: the use of indicators, GIS and case studies is often helpful in this context, as outlined earlier in this book. Another need is often for a transparent way of examining and assessing choices. Techniques, such as role-play and simulation, and tools such as decision-support systems, may be useful for this purpose. Formal methods, which allow the opinions and concerns of different stakeholders to be brought together and compared, may also be required. Relevant methods include Delphi techniques (Richey *et al.*, 1985a,b; Jones and Hunter, 1995) and multi-criteria assessment methods (Janikowski, 1998).

All these methods can be incorporated readily into the HEADLAMP process. Other methods can also be developed locally, to suit the situation. In the long term, however, it has to be recognised that these methods may not be enough. If deeper-rooted change in attitudes to environmental health and more far-reaching changes in health protection are to be encouraged, then it will require changes in the way people are informed and trained about issues such as risk and environmental health, from a young age. This implies the need to incorporate some of the philosophy of the HEADLAMP approach, and information from such studies, into the school and college curriculum. This may be the real legacy of applying the HEADLAMP approach in many areas. Indeed, this must surely be the long-term aim of most people: to enhance the understanding of future generations about the links between environment and health, so that they can avoid the mistakes of the past and, both through their individual actions and by their collective decisions, build a better world for the future. That is the real aim and the real meaning of sustainable development to which the HEADLAMP approach aspires.

9.7   References

Bobak, M. and Leon, D.A. 1992 Air pollution and infant mortality in the Czech Republic, 1986–88. *The Lancet,* **340**, 1010–4.

Briggs, D.J. 1998 Coping with complexity in environmental health management and policy. In: D.J. Briggs, R, Stern and T.L. Tinker [Eds] *Environmental Health for All. Risk Assessment and Risk Communication for National Environmental Health Action Plans.* NATO Science Series 2. Environmental Security - Vol. 49. Kluwer, Dordrecht, 233–43.

de Hoogh, C. 1999 Estimating exposure to traffic-related pollution within a GIS environment. Unpublished PhD thesis, University College Northampton and University of Leicester.

Douglas, M. and Wildawsky, A. 1982 *Risk and Culture*. University of California Press, Berkley.

Dunn, C. and Kingham, S. 1996 Establishing links between air quality and health: searching for the impossible? *Social Science Medicine,* **42**(6), 831–41.

Economopolous, A.P. 1993 *Assessment of Sources of Air, Water and Land Pollution. A Guide to Rapid Source Inventory Techniques and their Use in Formulating Environmental Control Strategies, (2 volumes).* World Health Organization, Geneva.

Elliott, P., Briggs, D., Lebret, E., Gorynski, P. and Kriz, B. 1995 Small Area Variations in Air Quality and Health, The SAVIAH study: Design and methods. (Abstract). *Epidemiology,* **6**(4), S32.

FAO/WHO 1989 *Summary of Acceptances: Worldwide and Regional Codex Standards. Codex Alimentarius Part 1, Rev. 4.* Food and Agriculture Organization of the United Nations/World Health Organization, Geneva.

Farago, K. 1998 Reality versus perception, and values versus science in risk assessment and risk perception. In: D.J. Briggs, R, Stern and T.L. Tinker [Eds] *Environmental Health for All. Risk Assessment and Risk Communication for National Environmental Health Action Plans.* NATO Science Series 2. Environmental Security - Vol. 49. Kluwer, Dordrecht, 223–32.

Forastiere, F., Corbo, G.M., Michelozzi, P., Pistelli, R., Agabiti, N., Brancato, G., Ciappi, G. and Peruicci, C.A. 1992 Effects of environment and passive smoking on the respiratory health of children. *International Journal of Epidemiology,* **21**, 66–73.

Goren, A.I. and Hellmann, S. 1988 Prevalence of respiratory symptoms and diseases in schoolchildren living in a polluted and in a low polluted area in Israel. *Environmental Research*, **45**, 28–37.

Goren, A.I., Hellmann, S., Brenner, S., Egoz, N. and Rishpon, S. 1990 Prevalence of respiratory conditions among schoolchildren exposed to different levels of air pollutants in the Haifa Bay area, Israel. *Environmental Health Perspectives*, **89**, 225–31.

Jaakola, J.J., Paunio, M., Virtanen, M. and Heinonen, O.P. 1991 Low-level air pollution and upper respiratory infections in children. *American Journal of Public Health*, **81**, 1060–3.

Janikowski, R. 1998 Priority setting of environmental and health policy options. In: D.J. Briggs, R. Stern and T.L. Tinker [Eds] *Environmental Health for All. Risk Assessment and Risk Communication for National Environmental Health Action Plans.* NATO Science Series 2. Environmental Security - Vol. 49. Kluwer, Dordrecht, 175–85.

Jardine, C.G., Krahn, H. and Hrudey, S.E. 1995 Health risk perception in Alberta. Unpublished, Eco-Research Chair in Environmental Risk Management, Research report 95-1, University of Alberta.

Jardine, C. and Hrudey, S.E. 1998a What is risk? In: D.J. Briggs, R. Stern and T.L. Tinker [Eds] *Environmental Health for All. Risk Assessment and Risk Communication for National Environmental Health Action Plans.* NATO Science Series 2. Environmental Security - Vol. 49. Kluwer, Dordrecht, 205–11.

Jardine, C. and Hrudey, S.E. 1998b Promoting active public participation. In: D.J. Briggs, R. Stern and T.L. Tinker [Eds] *Environmental Health for All. Risk Assessment and Risk Communication for National Environmental Health Action Plans.* NATO Science Series 2. Environmental Security - Vol. 49. Kluwer, Dordrecht, 157–68.

Jones, J. and Hunter, D. 1995 Consensus methods for medical and health services research. *British Medical Journal,* **311**, 376–80.

Pages, J.P., Brenot, J., Bastide, S. and Carde, C. 1988 Risk perception, conflicts and decision. Unpublished paper presented at IIASA Conference on Risk, Laxenburg, Austria.

Powell, D. and Leiss, W. 1997 *Mad Cows and Mother's Milk: the Perils of Poor Risk Communication.* McGill-Queen's University Press, Kingston.

Presidential/Congressional Commission on Risk Assessment and Risk Management 1997 Framework for Environmental Health Risk Management. Final report, Volume 1, Presidential/Congressional Commission on Risk Assessment and Risk Management, Washington.

Richey, J.S., Mar, B.W. and Horner, R.R. 1985a The Delphi technique in environmental assessment - I. Implementation and effectiveness. *Journal of Environmental Management*, **21**, 135–46.

Richey, J.S., Horner, R.R. and Mar, B.W. 1985b The Delphi technique in environmental assessment - II. Consensus on critical issues in environmental monitoring program design. *Journal of Environmental Management*, **21**, 147–59.

Rivett, P. 1994 *The Craft of Decision Modelling.* John Wiley & Sons, Chichester.

Taubes, G. 1995 Epidemiology faces its limits. *Science*, **269**,164–9.

U.S. National Academy of Sciences 1991 *Human Exposure Assessment for Airborne Pollutants. Advances and Opportunities.* National Academy of Sciences, National Academy Press, Washington, D.C., 321 pp.

U.S. National Research Council 1996 *Understanding Risk: Informing Decisions in a Democratic Society.* National Academy Press, Washington D.C.

WHO 1980 *Recommended Health-Based Limits in Occupational Exposure to Heavy Metals.* WHO Technical Report Series 647, World Health Organization, Geneva.

WHO 1982 Rapid assessment of sources of air, water, and land pollution. WHO offset publication No. 62, World Health Organization, Geneva.

WHO 1986 *Recommended Health-Based Limits in Occupational Exposure to Selected Mineral Dusts.* WHO Technical Report Series 734, World Health Organization, Geneva.

WHO 1987 *Air Quality Guidelines for Europe.* WHO Regional publications, European Series No. 23, World Health Organization, Copenhagen.

WHO 1993 *Guidelines for Drinking-Water Quality. Vol. 1. Recommendations.* World Health Organization, Geneva.

Wills, J. 1998 The development and use of environmental health indicators for epidemiology and policy applications: a geographical analysis. Unpublished PhD thesis, University College Northampton and University of Leicester.

Annex 1

EXAMPLES OF ENVIRONMENTAL HEALTH INDICATORS

Table A	**Population in informal settlements**
Table B	**Population living in unsafe housing**
Table C	**Accidents in the home**
Table D	**Urban planning and building regulations**
Table E	**Access to basic sanitation**
Table F	**Diarrhoea morbidity in children**
Table G	**Diarrhoea mortality in children**
Table H	**Ambient concentrations of air pollutants in urban areas**
Table I	**Sources of indoor air pollution**
Table J	**Childhood morbidity due to acute respiratory illness**
Table K	**Childhood mortality due to acute respiratory illness**
Table L	**Capability for air quality management**
Table M	**Availability of lead-free petrol**

Table A

Population in informal settlements	An exposure indicator

Indicator profile

Issue   Shelter

Rationale and role   Rapid urbanisation, and inadequate capability to cope with the housing needs of people in urban areas, have contributed to the development of informal settlements. Living in these settlements often poses significant health risks: sanitation and drinking water quality are often poor with the result that inhabitants are exposed to a wide range of pathogens; cooking facilities are often basic with the consequence that high levels of exposure to indoor pollution may occur; and access to health and other services may be limited.

This indicator thus provides a measure of exposure to inadequate housing conditions. It can be used to:

- compare areas or countries in terms of their extent of informal settlements and the adequacy of their housing;
- monitor trends in the extent of informal settlements (e.g. in response to urbanisation or population change);
- identify areas characterised by poor housing conditions in order to target action;
- help investigate associations between housing conditions and health;
- assess and monitor the effectiveness of interventions aimed at improving housing conditions.

Linkage with other indicators   This indicator is one of a chain of indicators describing the health risks associated with inadequate shelter. Others are:

- Exposure: **Population in informal settlements**; *Population living in unsafe housing.*
- Effect: *Accidents in the home.*
- Action: *Urban planning and building regulations.*

However, the characteristics of, and health risks associated with, living in informal settlements extend more widely than this, and other relevant indicators include:

- Exposure: *Access to basic sanitation; Connections to piped water; Access to safe and reliable supplies of drinking water.*
- Effect: *Diarrhoea morbidity in children; Diarrhoea mortality in children; Childhood morbidity due to acute respiratory illness; Childhood mortality due to acute respiratory illness.*

Continued

Table A   Continued

Population in informal settlements	An exposure indicator

Alternative methods and definitions	This indicator can be defined as the percentage of the population living in informal settlements. It is often restricted to informal settlements within the urban environment, and as such may omit contiguous peri-urban settlements. An urban focus makes the indicator less comprehensive, but data are likely to be more readily available, and of better quality, than for rural settlements. The indicator might also be presented as the total number of people living in informal settlements.
	Where suitable data on population are not available, the indicator might alternatively be measured as the area (e.g. in km^2) of informal settlements. This may be estimated from aerial photographs. It is liable to understate the scale of the problem, however, because it makes no allowance for population density, which is often higher in informal settlements than in formal settlements.
	Other measures of marginal human settlements have been formulated, many of which could be used to develop similar indicators. These include unplanned settlements, squatter settlements, marginal settlements, unconventional dwellings, non-permanent structures, inadequate housing, slums and housing in compliance.
	"Unconventional dwellings" are commonly defined by the number of housing units occupied by households, but considered inappropriate to human habitation.
	"Housing in compliance" is used as a human settlements indicator by the UN Habitat Programme and is defined as the percentage of the total housing stock in urban areas which is in compliance with current regulations (i.e. authorised housing).
	Housing may also be categorised by its type or permanence (e.g. permanent, semi-permanent, non-permanent), although definitions of these categories vary widely from country to country.
Related indicator sets	UN Indicators of Sustainable Development: • Area and population of urban formal and informal settlements.
Sources of further information	UN 1996 *Indicators of Sustainable Development: Framework and Methodologies.* United Nations, New York.
	UNCHS (Habitat)/World Bank 1993 *The Housing Indicators Programme.* Report of the Executive Director (Volume I). United Nations Centre for Human Settlements (Habitat), Nairobi.

Continued

Table A   Continued

Population in informal settlements   An exposure indicator

	UNCHS (Habitat) 1995 *Monitoring the Shelter Sector.* Housing Indicators Review. United Nations Centre for Human Settlements (Habitat), Nairobi.
	UNCHS (Habitat) 1995 *Monitoring Human Settlements: Abridged Survey.* Indicators Programme. United Nations Centre for Human Settlements (Habitat), Nairobi.
	UNCHS (Habitat) Urban Indicators Programme web page: http://www.urbanobservatory.org/indicators/database/.
Involved agencies	UN Centre for Human Settlements (Habitat)
	The World Bank
	WHO — Healthy Cities Project

Example indicator

Definition of indicator	Percentage of the population living in informal settlements.
Underlying definitions and concepts	This indicator depends on the ability to define and measure the number of people living in informal settlements. It covers both urban and rural settlements. Underlying definitions are:
	"Informal settlements": various definitions have previously been proposed, but that suggested by the UN Habitat Programme is probably the most appropriate. This defines informal settlements as:
	i) residential areas where a group of housing units has been constructed on land to which the occupants have no legal claim or which they occupy illegally; and
	ii) unplanned settlements and areas where housing is not in compliance with current planning and building regulations (i.e. unauthorised housing).
	"Unauthorised housing": excludes units where land titles, leases or occupancy permits have been granted (UN, 1996).
	"Total population": total resident population.
	It should be noted that informal settlements do not cover the homeless.
Specification of data needed	Number of people living in informal settlements. Total population.
Data sources, availability and quality	Data on the number of people living in informal settlements are often limited, because inhabitants are often only inadequately covered by formal censuses; census data may therefore not

Continued

Table A   Continued

Population in informal settlements	An exposure indicator

provide a clear separation of those living in informal settlements. Where suitable census data do not exist, special surveys may be necessary.

Data on total population should be available from national censuses and are generally reliable (except for those living in informal settlements).

Computation	The indicator is computed as:

$$100 * (P_i / P_t)$$

where P_i is the population living in informal settlements and P_t is the total population.

Units of measurement	Percentage.

Interpretation	This indicator provides a relatively straightforward measure of the quality of housing. A large percentage of people living in informal settlements can be taken to imply relatively poor housing conditions; a low percentage implies better housing conditions.

Nevertheless, the relationship between the number of people living in informal settlements and environmental health is not always simple. In particular, those living in formal settlements are not necessarily better provided for (e.g. the homeless or people living in crowded or unsafe housing). Problems of data accuracy also mean that the indicator should be interpreted with care, especially where comparisons are being made between different surveys.

Table B

Population living in unsafe housing	An exposure indicator

Indicator profile

Issue | Shelter

Rationale and role | The adequacy of housing is an important determinant of health status in a number of ways. *Inter alia*, housing quality affects levels of exposure to indoor pollutants, food and water hygiene, levels of sanitation, exposures to physical hazards and injury, and the general quality of life.

Housing may be unsafe for a variety of reasons, including dangerous construction, inadequate ventilation, inadequate heating, dangerous or inadequately maintained services, inadequate size for the number of residents (i.e. overcrowding) or location in a hazardous area (e.g. in areas prone to flooding or earthquakes, or on contaminated land). Living in inadequate housing is therefore likely to result in increased risks of a variety of health effects, including respiratory illness, gastro-intestinal infections and infant mortality.

This indicator provides a general measure of the adequacy of the housing stock, and the level of exposures to those hazards which might thus occur. Potential uses include:
- monitoring the general adequacy of the housing stock, and access to this stock by the population;
- monitoring the magnitude and implications of major demographic or social changes in the population (e.g. as a result of rapid urbanisation or migration);
- assessment of changes in the general level of health risk associated with poor housing;
- mapping risks associated with poor housing, in order to identify areas of special need;
- assessing the effectiveness of national or regional strategies aimed at improving the housing stock;
- analysing relationships between quality of housing and health effects.

Linkage with other indicators | This indicator is part of a chain of indicators which collectively describe the risks associated with inadequacy of shelter:
- Exposure: *Population in informal settlements;* **Population living in unsafe housing.**
- Effect: *Accidents in the home.*
- Action: *Urban planning and building regulations.*

Continued

Table B   Continued

Population living in unsafe housing	An exposure indicator

	However, the characteristics of, and health risks associated with, unsafe, unhealthy or hazardous housing extend more widely than this, and other relevant indicators include: • Exposure: *Access to basic sanitation; Connection to piped water; Access to safe and reliable supplies of drinking water.* • Effect: *Diarrhoea morbidity in children; Diarrhoea mortality in children; Childhood morbidity due to acute respiratory illness; Childhood mortality due to acute respiratory illness; Outbreaks of waterborne diseases.*
Alternative methods and definitions	Although potentially valuable, this indicator is difficult to define and measure in a clear and systematic manner. The most appropriate measure would be the percentage (or number) of people living in unsafe, unhealthy or hazardous housing. However, defining the terms "unsafe", "unhealthy" and "hazardous" poses severe difficulties, as does obtaining data on houses which meet these criteria. A somewhat weaker alternative to this indicator can be obtained by assessing the percentage of the total housing stock which is considered unsafe, unhealthy or hazardous. Information can be obtained from housing condition surveys. This is liable to underestimate the number of people affected because of the tendency for overcrowding in poorer quality housing. A further alternative is to use census derived data (e.g. on overcrowding or the availability of basic amenities in the home), where these exist, as a measure of inadequate housing. These terms are usually defined nationally by the census. Where the main concern is about natural hazards, such as flooding, earthquakes, avalanches or radon exposures, estimates of the exposed population may be made using GIS techniques to map hazardous areas and overlay these with population data.
Related indicator sets	UNCHS (Habitat) Urban Indicators Programme: • Permanent structures (percentage of housing units located in structures expected to maintain their stability for 20 years or longer under local conditions with normal maintenance). • Housing in compliance (percentage of the total housing stock in compliance with current regulations). • Housing destroyed (percentage of the housing stock destroyed by natural or man-made disasters over the past ten years).

Continued

Table B   Continued

Population living in unsafe housing   An exposure indicator

Sources of further information	WHO 1994 *Implementation of the Global Strategy for Health for All by the Year 2000. Second Evaluation. Eighth Report on the World Health Situation.* (Volume 5: European Region). WHO Regional Office for Europe, Geneva.
	WHO 1997 *Health and Environment in Sustainable Development: Five Years after the Earth Summit.* World Health Organization, Geneva.
	UNCHS (Habitat) Urban Indicators Programme web page: http://www.urbanobservatory.org/indicators/database/.
Involved agencies	UN Centre for Human Settlements (Habitat)
	WHO — Healthy Cities Programme
	National, regional and local housing agencies

Example Indicator

Definition of indicator	Percentage of the population living in unsafe, unhealthy or hazardous housing.
Underlying definitions and concepts	This indicator requires the ability to identify, and measure the extent of, unsafe, unhealthy or hazardous housing. This poses significant difficulties, because these are all to a large extent both environmentally and culturally dependent, and thus are liable to vary from one area (or one time) to another. Possible definitions of unsafe, unhealthy or hazardous housing include housing which is:

- physically unsound and likely to be dangerous to its occupants because of its poor construction, or inadequately maintained services (e.g. electricity); or
- is located in a physically hazardous area (e.g. in an area of flood or earthquake risk) or is sited on contaminated land (e.g. by chemical wastes or radioactivity); or
- provides serious risks of exposures to indoor pollution (e.g. air pollutants) or pathogens (e.g. moulds, ticks and fleas); or
- provides inadequate shelter (e.g. due to poor insulation or inadequate roofing) and few or no basic amenities (e.g. cooking facilities and heating).

In addition, a definition is required of the total population (i.e. the total resident population at the time of the census or survey).

Continued

Table B   Continued

Population living in unsafe housing	An exposure indicator

Specification of data needed	Number of people living in unsafe, unhealthy or hazardous housing. Total resident population.
Data sources, availability and quality	Data on the quality of the housing stock, and the number of people living in unsafe, unhealthy or hazardous housing is rarely available from routine sources. In some countries, an approximation to this may be available from census statistics (e.g. housing lacking basic amenities). Generally, however, data will need to be obtained by special surveys. In all cases, these data are liable to considerable margins of error and inconsistency due to difficulties of definition, inconsistent reporting and difficulties of ensuring representative sampling. Data on the total resident population should be available from national censuses and should be reliable.
Computation	The indicator can be computed as: $100 * (P_u / P_t)$ where P_u is the number of people living in unsafe, unhealthy or hazardous housing and P_t is the total population.
Units of measurement	Percentage.
Interpretation	This is an important indicator which has wide-ranging significance for policy. In providing a measure of the adequacy of the housing stock, it also acts as an indicator of health risks associated with poor sanitation, exposures to indoor air pollution and access to safe water. It can therefore help to interpret a range of other issues and indicators. Like all general purpose indicators, however, it needs to be interpreted carefully. The characteristics which render housing unsafe, unhealthy or hazardous may clearly vary; without information on these specific characteristics it can be misleading to infer either the existence of particular health risks or effects or the need for specific actions. Definitional issues are also likely to pose major difficulties for comparisons between different areas, or between different surveys, unless standard protocols have been used. A clear understanding of the data is therefore essential before interpretations are made.

Table C

Accidents in the home	An effect indicator

Indicator profile

Issue	Shelter
Rationale and role	Accidents in the home are one of the main causes of injury and death. Although accidents can occur in any home, the risk of accidents tends to be increased by poor building design and inadequate safety requirements for housing. This indicator thus provides a measure of the effect of inadequate housing. It can be used:

- to monitor the incidence of accidents in the home;
- to identify areas or types of housing with unacceptably high rates of accident or injury, as a basis for targeting action;
- to help develop and design safer houses;
- to help establish more effective planning and building regulations;
- to assess the effectiveness of policy interventions, aimed at reducing accidents in the home (e.g. new building regulations or awareness raising campaigns).

Linkage with other indicators	This indicator is part of a chain of indicators that collectively describe the risks associated with inadequacy of shelter:

- Exposure: *Population in informal settlements; Population living in unsafe housing.*
- Effect: **Accidents in the home.**
- Action: *Urban planning and building regulations.*

Alternative methods and definitions	This indicator can be defined as the incidence of injury by accidents in the home. Because the young and elderly are the most vulnerable to accidents in the home, it may be appropriate to stratify the indicator by age (and perhaps gender) or to restrict it to specific age groups.
Related indicator sets	None.
Sources of further information	
Involved agencies	WHO

Continued

Table C   Continued

Accidents in the home	An effect indicator

Example indicator

Definition of indicator	Incidence of injury by accidents in the home.
Underlying definitions and concepts	"Accidents in the home": an accident, taking place in the home, which leads to physical injury sufficient to require medical treatment. Common accidents include falling down stairs, electrocution, burning, scalding and accidents with kitchen utensils and equipment. For the purpose of this indicator, poisonings should be excluded, if possible.
	"Total population": total resident population.
Specification of data needed	Number of reported accidents in the home. Total population.
Data sources, availability and quality	Comprehensive data on physical injuries by accidents in the home are likely to be difficult to acquire, due to lack of referral or reporting. Many injuries may not be considered sufficient to be referred to the medical services; many others, although reported, may not be clearly classified as a result of an accident in the home. Probably the most useful source of data are hospital admission statistics, although these tend to cover the more severe, acute injuries. Other potential sources include data from GPs and household surveys.
	Data on the total population should be available from national census statistics, and should be reliable.
Computation	The indicator can be computed as:
	$1,000 * (A / P)$
	where A is the total number of reported cases of injury by accidents in the home, and P is the total population.
Units of measurement	Number per thousand head of population.
Interpretation	This is a potentially useful indicator, which gives a general measure of injuries due to accidents in the home.
	Problems of data availability and quality, however, mean that care is needed in making comparisons between different areas or countries, or over long periods of time. Data are likely to be affected, for example, by ease of access to the medical services, and by differences in reporting procedures.

Table D

Urban planning and building regulations	An action indicator

Indicator profile

Issue	Shelter
Rationale and role	The application of strict building and planning regulations for housing is one of the main ways by which health risks of inadequate housing can be mitigated. Such regulations can control development on unsuitable sites (e.g. contaminated, unstable or flood-prone land), and set minimum standards for residential accommodation (e.g. in terms of space, quality of construction and safety). This indicator is thus an action indicator, aimed at assessing the level of commitment made to ensuring safe housing. It is relevant mainly at the international level, for example, to:

- compare countries in terms of their level of planning and building regulations;
- monitor national trends towards the establishment of adequate planning and building control;
- help interpret inter-country variations in the quality of housing and levels of morbidity and mortality relating to inadequate housing.

Linkage with other indicators	This indicator is part of a chain of indicators which collectively describe the risks associated with inadequacy of shelter:

- Exposure: *Population in informal settlements; Population living in unsafe housing.*
- Effect: *Accidents in the home.*
- Action: **Urban planning and building regulations.**

Alternative methods and definitions	Like most indicators relating to the effectiveness or adequacy of policy and management, this indicator is not easy to define and apply in a stringent and systematic way. Possibly the best that can normally be achieved is to assess the existence and rigour of building and planning regulations for residential housing (see example below). It needs to be recognised, however, that the existence of such regulations does not necessarily mean that they are being effectively applied. Alternatively, the indicator could be assessed in terms of the proportion of the housing stock covered by formal building regulations.
	More complex indicators could be developed by defining in more detail the elements of building regulations and planning consents and, if appropriate, by separating the regulations relating to public and private housing development.

Continued

Table D   Continued

Urban planning and building regulations	An action indicator

Related indicator sets   None.

Sources of further information   UNCHS (Habitat)/World Bank 1993 *The Housing Indicators Programme.* Report of the Executive Director (Volume I). United Nations Centre for Human Settlements (Habitat), Nairobi.

UNCHS (Habitat) 1995 *Monitoring the Shelter Sector.* Housing Indicators Review. United Nations Centre for Human Settlements (Habitat), Nairobi.

UNCHS (Habitat) 1995 *Monitoring Human Settlements: Abridged Survey.* Indicators Programme. United Nations Centre for Human Settlements, Nairobi.

UNCHS (Habitat) 1998 *People, Settlements, Environment and Development.* United Nations Centre for Human Settlements (Habitat), Nairobi.

UNCHS (Habitat) Urban Indicators Programme web page: http://www.urbanobservatory.org/indicators/.

Involved agencies   UN Centre for Human Settlements (Habitat)

WHO

Example indicator

Definition of indicator   Scope and extent of building regulations for housing.

Underlying definitions and concepts   This indicator is based on the assumption that urban planning and building regulations can help to reduce health risks by controlling residential development on unsuitable sites and by providing adequate standards for housing construction and design. Underlying definitions are:

"Land use planning": formal procedures for controlling where, and under what conditions, land is developed for housing and other purposes. These procedures usually require formal consent before development and construction can occur. Land may also be zoned, with specific areas designated for housing purposes.

"Building regulations": legally defined standards and norms for building which must be met by the developer. Building regulations may cover issues such as the amount of space per occupant, construction materials and methods, and safety standards.

Continued

Table D   Continued

Urban planning and building regulations	An action indicator

Specification of data needed	Evidence of the existence, implementation and enforcement of land use planning and building regulations for housing.
Data sources, availability and quality	Evidence can normally best be obtained by scrutinising relevant legislation.
Computation	The indicator is computed by scoring 1 for each of the following components: • formal planning consent required for all residential development; • strict land zoning in existence that defines areas suitable/permissible for housing; • building regulations exist that define minimum space requirements and living conditions (e.g. lighting and insulation) for houses; • building regulations exist that control building methods and materials for houses; • building regulations exist that define safety standards for houses.
Units of measurement	Ordinal score (0–5).
Interpretation	This indicator provides a general measure of the rigour and scope of building and planning regulations for housing, and thus of the level of commitment to ensuring safe and adequate housing. The simple scoring system, however, means that it should be interpreted with caution, not least because the existence of the various regulations and planning instruments does not necessarily mean that they are effectively implemented and enforced.

Table E

Access to basic sanitation	An exposure indicator

Indicator profile

Issue	Sanitation
Rationale and role	Access to adequate excreta disposal facilities is an important requirement if adverse health effects of poor sanitation are to be avoided. This indicator thus provides a measure both of the potential exposure of the population to infectious agents associated with poor sanitation, and of the action taken to improve domestic sanitation. The indicator can be used:

* to assess and compare general levels of access to sanitation facilities as a basis for priority setting;
* as one of a group of indicators to assess levels of social inequality and deprivation;
* to assess and identify areas with poor sanitation, where specific policy action may be required;
* to help investigate associations between sanitary conditions and specific health effects;
* to help target and plan efforts to improve domestic sanitation and to monitor progress of such measures.

Linkage with other indicators	This indicator is part of a chain of indicators, collectively describing the effects on health of access to basic sanitation, water quality and access, and food safety:

* Exposure: *Access to basic sanitation*.
* Effect: *Diarrhoea morbidity in children; Diarrhoea mortality in children*.

Alternative methods and definitions	The indicator can be defined as the percentage of the population (or of households) with (or alternatively without) access to adequate excreta disposal facilities. To apply this definition, a clear and appropriate definition is needed of what constitutes "adequate excreta disposal facilities". This needs to specify both the type of facility and its accessibility (e.g. whether in the home or outside). Definitions are likely to vary according to local circumstances (e.g. between developed and developing countries).
	Where data are available, the indicator could be further refined according to the type of facilities (e.g. connection to public sewage system, cess-pit, pit latrines and facilities in house or outside).

Continued

Table E   Continued

Access to basic sanitation   An exposure indicator

Related indicator sets	UN Indicators of Sustainable Development: • Basic sanitation: percent of population with adequate excreta disposal facilities. WHO Catalogue of Health Indicators: • Access to sanitary means of excreta disposal.
Sources of further information	UN 1996 *Indicators of Sustainable Development: Framework and Methodologies.* United Nations, New York. WHO 1981 *Development of Indicators for Monitoring Health for All by the Year 2000.* World Health Organization, Geneva. WHO 1982 *National and Global Monitoring of Water Supply and Sanitation.* VWS Series of Cooperative Action for the Decade, No.2. World Health Organization, Geneva. WHO 1990 *Water Supply and Sanitation Sector Monitoring Report (WSSMR).* WHO/UNICEF Joint Monitoring Programme. World Health Organization, Geneva. WHO 1994 *Ninth General Programme of Work Covering the Period 1996–2001.* World Health Organization, Geneva. WHO 1996 *Catalogue of Health Indicators: A Selection of Health Indicators Recommended by WHO Programmes.* World Health Organization, Geneva.
Involved agencies	WHO — Programme for the Promotion of Environmental Health

Example indicator

Definition of indicator	Percentage of the population with access to adequate excreta disposal facilities.
Underlying definitions and concepts	This indicator is based on the assumption that poor sanitary facilities increase the risks of infectious diseases such as diarrhoea and cholera. Underlying definitions are: "Adequate excreta disposal facilities": a facility which provides for the controlled disposal of human excreta in ways which avoid direct human exposure to faeces, or contamination of food and local water supplies by raw faeces. Suitable facilities might range from simple but effective pit latrines, to flush toilets with sewerage. All facilities, to be effective, must be correctly constructed and properly maintained.

Continued

Table E   Continued

Access to basic sanitation	An exposure indicator

	"Access to adequate excreta disposal facilities": people with excreta disposal facilities either in their dwelling or located within a convenient distance (< 50 metres) from the user's dwelling. This thus includes the urban and rural populations served by connections to public sewers, household systems (e.g. pit privies, pour-flush latrines and septic tanks), communal toilets and simple but adequate excreta disposal such as pit privies, pour-flush latrines and covered by latrines.
	"Total population": total resident population.
Specification of data needed	The number of people with access to adequate excreta disposal facilities. Total population.
Data sources, availability and quality	Data on excreta disposal facilities may be available from relevant administrative authorities (e.g. public works, sanitary works or housing departments). In some countries, data are also available via national censuses. Where such sources do not exist, or are inadequate, special surveys will be necessary. Data on total population are available from national censuses and should be reliable.
Computation	The indicator can be computed as: $100 * (P_e / P_t)$ where P_e is the number of people living in dwellings with access to adequate excreta disposal facilities, and P_t is the total population.
Units of measurement	Percentage.
Interpretation	The indicator can be interpreted directly to show the adequacy of domestic sanitary conditions, and thus the risks to health from exposures to infectious agents. A high percentage of people or households with access to adequate excreta disposal facilities should indicate a lower risk of exposure and adverse health effects; a low percentage would imply higher risks of exposure and infection. If compared with national targets, the indicator can similarly be interpreted to show progress towards achieving these goals. Nevertheless, some care is needed in interpreting the indicator, in particular because the availability of a facility does not always translate into their proper use and improved hygiene. Data may also be of uncertain quality.

Table F

Diarrhoea morbidity in children	An effect indicator

Indicator profile

Issue

Sanitation

Access to safe drinking water

Food safety and supply

Rationale and role

This indicator measures the health effects of diarrhoea in the high risk group of under five-year olds. It is an indication of the magnitude of the problem of diarrhoea and the potential health effects from exposure to the environmental problems of poor quality sanitation, water and food.

As a measurement of cause-specific morbidity, this indicator can serve several purposes:

- to establish the magnitude of the problem of childhood diarrhoea and its relative public health importance;
- to evaluate trends over time, especially as a method of evaluating the probable impact of intervention, management and control programmes;
- to select, place and programme interventions;
- to provide an indication of the potential for health effects associated with the same environmental health issues.

Linkage with other indicators

This indicator is part of a number of chains of indicators, collectively describing the effects on health of access to basic sanitation, water quality and access, and food safety.

1. Sanitation
- Exposure: *Access to basic sanitation.*
- Effect: *Diarrhoea mortality in children;* **Diarrhoea morbidity in children.**

2. Access to safe drinking water
- Exposure: *Connections to piped water supply; Access to safe and reliable supplies of drinking water.*
- Effect: **Diarrhoea morbidity in children**; *Diarrhoea mortality in children; Outbreaks of waterborne diseases.*
- Action: *Intensity of water quality monitoring (specifically for drinking waters).*

3. Food safety and supply
- Effect: *Food-borne illness;* **Diarrhoea morbidity in children**; *Diarrhoea mortality in children.*
- Action: *Monitoring of chemical hazards in food.*

Continued

Table F   Continued

Diarrhoea morbidity in children	An effect indicator

Alternative methods and definitions	This indicator can be defined as the incidence of diarrhoea in children under five years of age. Where appropriate it could be applied to other age groups (e.g. 0–1 year olds). Alternatively, the indicator can be assessed on the basis of the number of hospital admissions for acute gastro-intestinal infections. This, however, would tend to underestimate the incidence of illness because only the most serious cases would be included. Bias might also occur in the indicator, because of social and geographic differences in access to hospitals.
Related indicator sets	WHO Catalogue of Health Indicators: • Annual incidence of diarrhoea in children under five years of age.
Sources of further information	WHO 1992 *Readings on Diarrhoea: Student Manual.* Division for the Control of Diarrhoea and Acute Respiratory Disease, World Health Organization, Geneva. WHO 1994 *Ninth General Programme of Work Covering the Period 1996–2001.* World Health Organization, Geneva. WHO 1994 *Household Survey Manual: Diarrhoea and Acute Respiratory Infections.* WHO/CDR/94.8. World Health Organization, Geneva. WHO 1996 *Catalogue of Health Indicators: A Selection of Health Indicators Recommended by WHO Programmes.* World Health Organization, Geneva. WHO 1997 *Health and Environment in Sustainable Development: Five Years After the Earth Summit.* World Health Organization, Geneva.
Involved agencies	WHO — Department of Child and Adolescent Health and Development (CAH) WHO — Programme for the Promotion of Environmental Health UNICEF

Example indicator

Definition of indicator	Incidence of diarrhoea morbidity in children under five years of age.
Underlying definitions and concepts	"Diarrhoea": three or more watery stools in a 24-hour period, a loose stool being one that would take the shape of the container (WHO, 1996), or local definition of diarrhoea.

Continued

Table F   Continued

Diarrhoea morbidity in children   An effect indicator

"Episode of diarrhoea": an episode of diarrhoea begins with a 24-hour period with three or more loose or watery stools. An episode of diarrhoea is considered to have ended after 48 hours without three or more loose watery stools within a 24-hour period.

"Incidence of diarrhoea morbidity": the total number of episodes of diarrhoea during a one-year period amongst the children surveyed.

"Total population of children under five years of age": the number of children less than five years of age in the survey, at the time of the survey.

Specification of data needed	Data on the number of episodes of diarrhoea among children under five years of age. Population data for the total number of children under five years of age. Disaggregating data such as socio-economic status, geographic area and age/sex of children.
Data sources, availability and quality	Morbidity data for diarrhoeal disease do not tend to be collected on a routine basis, and usually depend on special surveys. Methods for data collection by surveys are recommended by the WHO Division for the Control of Diarrhoea and Acute Respiratory Disease (CDD/ARI) household survey manual (see Sources of further information). The CDD/ARI Household Survey is designed to collect qualitative as well as quantitative information on diarrhoea episodes occurring in the past two weeks. The manual includes instructions on how to convert the results to an annual incidence taking into account seasonal variations.
Computation	The indicator can be computed as: I_c / P_c where I_c is the incidence of diarrhoea in children under five years of age in the survey, and P_c is the total number of children under five years of age in the survey.
Units of measurement	Number of cases per child per year.

Continued

Table F   Continued

Diarrhoea morbidity in children   An effect indicator

Interpretation   This indicator is a powerful measure of the health status of
                        children, especially under conditions of inadequate water and
                        food hygiene, and poor basic sanitation. Action to improve these
                        conditions can generally help to reduce morbidity rates. Like
                        other infectious diseases, however, marked short-term
                        variations in morbidity may occur, making identification of
                        long-term trends difficult, especially on the basis of short-term
                        or irregular surveys. Data on the incidence of diarrhoea are
                        also subject to large margins of error due to inconsistencies in
                        reporting and in definitions, and problems of ensuring adequate
                        sampling in surveys.

                        Interpretation of the indicator can be assisted by disaggregating
                        the data by age and gender of the child, economic status of the
                        parents and by geographic area.

Table G

Diarrhoea mortality in children	An effect indicator

Indicator profile

Issue	Sanitation
	Access to safe drinking water
	Food safety and supply
Rationale and role	Diarrhoea and related gastro-intestinal illnesses continue to be among the most important causes of illness and death worldwide, especially amongst vulnerable groups such as young children. Much of this illness is due to exposure to contaminated water or food, as a result, for example, of poor water quality, limited access to water, poor food hygiene and safety, or poor sanitation in the home. Major pathogens include *Salmonella*, *Shigella*, *Campylobacter*, *E. coli* and rotavirus.

This indicator provides a measure of the extent and severity of these effects. It can thus be used:

- to monitor general trends in the burden of disease amongst children;
- to infer changes in the quality of drinking and bathing water, food and basic sanitation;
- to map patterns of disease as a basis for identifying at-risk areas or groups and to target policy action;
- to assess and monitor the effectiveness of intervention programmes;
- to analyse relationships between environmental exposures and health.

Linkage with other indicators	This indicator is part of a number of chains of indicators, collectively describing the effects on health of access to basic sanitation, water quality and access, and food safety.

1. Sanitation

- Exposure: *Access to basic sanitation.*
- Effect: **Diarrhoea mortality in children**; *Diarrhoea morbidity in children.*

2. Access to safe drinking water

- Exposure: *Connections to piped water supply; Access to safe and reliable supplies of drinking water.*
- Effect: *Diarrhoea morbidity in children;* **Diarrhoea mortality in children**; *Outbreaks of waterborne diseases.*
- Action: *Intensity of water quality monitoring (specifically for drinking waters).*

Continued

Table G   Continued

Diarrhoea mortality in children	An effect indicator

3. Food safety and supply
- Effect: *Food-borne illness; Diarrhoea morbidity in children;* **Diarrhoea mortality in children***.*
- Action: *Monitoring of chemical hazards in food.*

Alternative methods and definitions	This indicator can be defined as the mortality rate due to diarrhoea in children under five years of age. It could alternatively be assessed using a broader category of illnesses (e.g. diseases of the digestive system — ICD codes 520–579). While this would broaden the potential range of exposures of relevance, it would tend to reduce inconsistencies due to diagnosis. It could also be applied to other age groups (e.g. under one year) where appropriate. Stratification by gender may be useful in some cases.
Related indicator sets	WHO Catalogue of Health Indicators: • Deaths due to diarrhoea among infants and children under five years of age.
Sources of further information	WHO 1992 *Readings on Diarrhoea: Student Manual.* Division for the Control of Diarrhoea and Acute Respiratory Disease, World Health Organization, Geneva. WHO 1994 *Ninth General Programme of Work Covering the Period 1996–2001.* World Health Organization, Geneva. WHO 1994 *Household Survey Manual: Diarrhoea and Acute Respiratory Infections.* WHO/CDR/94.8. World Health Organization, Geneva. WHO 1996 *Catalogue of Health Indicators: A Selection of Health Indicators Recommended by WHO Programmes.* World Health Organization, Geneva. WHO 1997 *Health and Environment in Sustainable Development: Five Years After the Earth Summit.* World Health Organization, Geneva.
Involved agencies	WHO — Department of Child and Adolescent Health and Development (CAH) WHO — Programme for the Promotion of Environmental Health UNICEF

Continued

Table G   Continued

Diarrhoea mortality in children   An effect indicator

Example indicator

Definition of indicator	Diarrhoea mortality rate in children under five years of age.
Underlying definitions and concepts	"Death due to diarrhoea in children under five years of age": death in which diarrhoea is defined as a primary cause of a child of less than five years of age at the time of death.
	"Total population of children under five years of age": number of live children less than five years of age at the mid-point of the survey year (or other survey period).
Specification of data needed	Total number of deaths due to diarrhoea in children under five years of age.
	Total population of children under five years of age.
Data sources, availability and quality	Data on death due to diarrhoea in children under five years of age should be available through national or regional/local death statistics. Differences in both diagnosis and reporting practice may be significant in these data, especially where diarrhoea is one of a number of symptoms (e.g. associated with malnutrition). Where statistical data are not available from routine sources, special surveys will be necessary.
	Data on the total population of children under five years of age should usually be available through national censuses. Inter-census estimates can be made using vital registration data, or demographic models. Care is needed in applying a consistent and appropriate census date, especially where marked seasonal patterns in birth may occur.
Computation	The indicator can be computed as:
	$1,000 * (M_c / P_c)$
	where M_c is the total number of deaths amongst children under five years of age and P_c is the total population of children under five years of age.
Units of measurement	Number per thousand children under five years of age.

Continued

Table G   Continued

Diarrhoea mortality in children	An effect indicator
Interpretation	This indicator is a powerful measure of health status of children, especially under conditions of inadequate water or food hygiene and basic sanitation. Action to improve these conditions can generally help to reduce mortality rates. Like other infectious diseases, however, marked short-term variations in mortality may occur, making identification of long-term trends difficult. Death of young children due to diarrhoea may also be a result of several different, and often inter-related, exposures; attributing changes in mortality to any one of these without consideration of the others might be misleading. Rates of mortality are also fundamentally affected by the effectiveness of, and access to, the health service and levels of awareness amongst parents.

Table H

Ambient concentrations of air pollutants in urban areas	A state indicator

Indicator profile

Issue	Air pollution
Rationale and role	The purpose of this indicator is to measure overall air quality and the potential exposure of people to air pollutants of health concern. The indicator may be used:

- to monitor trends in air pollution as a basis for prioritising policy actions;
- to map levels of air pollution in order to identify hotspots or areas in need of special action;
- to help assess the number of people exposed to excess levels of air pollution;
- to monitor levels of compliance with air quality standards;
- to assess the effects of air quality policies;
- to help investigate associations between air pollution and health effects.

Linkage with other indicators	This indicator represents one in a chain of indicators that together describe the effects of air pollution on health:

- State: **Ambient concentrations of air pollutants in urban areas**.
- Exposure: *Sources of indoor air pollution.*
- Effect: *Childhood morbidity due to acute respiratory illness; Childhood mortality due to acute respiratory illness.*
- Action: *Capability for air quality management; Availability of lead-free petrol.*

Alternative methods and definitions	This indicator may be designed and constructed in a number of ways. Where monitored data are available, it might usefully be expressed in terms of mean annual or percentile concentrations of air pollutants with known health effects (for example, CO, particulates (PM_{10}, $PM_{2.5}$, SPM), black smoke, SO_2, NO_2, O_3, VOCs, benzene and lead) in the outdoor air in urban areas. Alternatively, the indicator might be expressed in terms of the number of days on which air quality guidelines or standards are exceeded (although in this case comparisons need to be made with care because of possible changes or differences in the guideline values).

Continued

Table H   Continued

Ambient concentrations of air pollutants in urban areas   A state indicator

---

	Where monitoring data are unavailable, estimates of pollution levels may be made using air pollution models. Dispersion models are, however, dependent on the availability of emissions data; where these are not available, surveys may be conducted using rapid source inventory techniques (Economopolous, 1993). Because of potential errors in the models or the input data, results from dispersion models should ideally be validated against monitored data.
Related indicator sets	UN Indicators of Sustainable Development: • Ambient concentrations of pollutants in urban areas.
Sources of further information	Economopolous, A.P. 1993 *Assessment of Sources of Air, Water and Land Pollution: A Guide to Rapid Source Inventory Techniques and Their Use in Formulating Environmental Control Strategies.* (Volume I and II). World Health Organization, Geneva. UN 1996 *Indicators of Sustainable Development: Framework and Methodologies.* Report for the UN Commission on Sustainable Development. United Nations, New York. WHO 1987 *Air Quality Guidelines for Europe.* WHO Regional Publications, European Series No. 23. World Health Organization, Geneva. (Updated in 1998; see http://www.who.int). WHO 1991 *Global Strategy for Health for All by the Year 2000.* World Health Organization, Geneva. WHO 1994 *Ninth General Programme of Work Covering the Period 1996–2001.* World Health Organization, Geneva. WHO 1998 *Healthy Cities Air Management Information System (AMIS).* AMIS v. 2.0. CDROM. World Health Organization, Geneva.
Involved agencies	WHO — Programme for the Promotion of Environmental Health National air quality monitoring networks WHO European Centre for Environment and Health European Environment Agency and Air Quality Topic Centre

Example indicator

Definition of indicator	Mean annual and percentile concentrations of CO, particulates (PM_{10}, $PM_{2.5}$, SPM), SO_2, NO_2, O_3 and lead in the outdoor air in urban areas.

Continued

Table H   Continued

Ambient concentrations of air pollutants in urban areas   A state indicator

---

Underlying definitions and concepts	This indicator is based on the assumption that outdoor levels of air pollution in urban areas represent a significant source of exposure and health risk.
	Underlying definitions are:
	"Mean annual concentration": mean concentration of the pollutant of concern, averaged over all hours of the year.
	"Percentile concentration": concentration of pollutant of concern exceeded in $100 - X\%$ of hours, where X is the percentile as defined by the relevant standards.
Specification of data needed	Mean annual and percentile concentrations for CO, PM_{10}, $PM_{2.5}$, SPM, SO_2, NO_2, O_3 and lead.
	Site location, site type (e.g. kerbside, intermediate or background), monitoring method (e.g. passive sampler or continuous monitor) and sampling frequency.
Data sources, availability and quality	Data on ambient air pollution concentrations can be obtained from national or local monitoring networks, using either continuous (fixed-site) monitors or passive samplers.
	In addition, a growing volume of data can be obtained from the WHO Healthy Cities Air Management Information System (AMIS).
Computation	The indicator can be presented as:
	• the mean annual concentration,
	• the relevant (e.g. 98th) percentile concentration,
	or otherwise as appropriate (e.g. number of days/hours in excess of air pollution standard).
Units of measurement	$\mu g\ m^{-3}$, ppm or ppb as appropriate, or percentage of days when standards/guideline values are exceeded.
Interpretation	This indicator can be used to interpret both spatial patterns and temporal trends in air pollution levels. In general terms, an increase in pollutant concentrations may be taken to suggest an increase in exposures and raised health risk; a reduction in pollution levels implies a decrease in exposures and a reduction in health risk. Interpretation is often aided by reference to the relevant air quality guidelines or standards (e.g. by assessing the number of days or hours during which the standards are exceeded).

Continued

Table H   Continued

Ambient concentrations of air pollutants in urban areas   A state indicator

Several factors nevertheless need to be taken into account in interpretation. One of the most important is the siting of the monitors. As a measure of exposure, data are generally most relevant where monitoring sites are located in residential or densely populated areas. Allowance also needs to be made for the detection limits, accuracy and comparability of the measurement methods. In particular, care needs to be taken when comparing data from different monitoring networks, due to the possibility of differences in sampling or measurement techniques. When used as a basis for assessing exposure, it is also important to recognise that actual exposures depend fundamentally upon indoor concentrations and time activity patterns of individuals. As with all exposure measures, relationships with health are also subject to considerable confounding, which should be strictly controlled for in epidemiological studies.

Table I

Sources of indoor air pollution	An exposure indicator

Indicator profile

Issue	Air pollution
Rationale and role	Indoor exposures to air pollution are an important factor in respiratory illness and mortality. Much of this exposure relates to the use of fuels such as wood, kerosene, coal or dung for cooking and heating. The indicator thus provides a measure of the potential exposure to air pollution from indoor sources. It can be used:

- to show time trends in levels of potential exposure;
- to provide an early indication of the effects of changes in domestic energy supplies on indoor exposures to air pollution;
- to show geographic variations in levels of potential exposure;
- to compare areas or countries in terms of potential exposures;
- to monitor the effects of intervention strategies aimed at reducing sources of indoor exposures due to cooking and heating fuels.

Linkage with other indicators	This indicator represents one in a chain of indicators that together describe the effects of air pollution on health:

- State: *Ambient concentrations of air pollutants in urban areas.*
- Exposure: ***Sources of indoor air pollution***.
- Effect: *Childhood morbidity due to acute respiratory illness; Childhood mortality due to acute respiratory illness.*
- Action: *Capability for air quality management; Availability of lead-free petrol.*

Alternative methods and definitions	This indicator can be computed as the number or proportion of households (or population) that rely on fuels such as coal, wood, dung and kerosene (or other high emission and poorly ventilated systems) for heating and cooking. Relevant data are often available from household surveys.
	Alternatively, the indicator could be defined as the percentage of households connected to electricity and gas supplies. Data on this may be available from censuses or from the utility companies. Another possible alternative would be to base the indicator on the percentage of total energy consumption provided by electricity or gas.
Related indicator sets	None.

Continued

Table I   Continued

Sources of indoor air pollution	An exposure indicator
Further sources of information	WHO 1994 *Implementation of the Global Strategy for Health for All by the Year 2000. Second Evaluation. Eighth Report on the World Health Situation.* (Volume 5: European Region). WHO Regional Office for Europe, Geneva
	WHO 1998 *Healthy Cities Air Management Information System (AMIS).* AMIS v. 2.0. CDROM. World Health Organization, Geneva.
Involved agencies	National energy supply companies
	National ministries of energy
	WHO

Example indicator

Definition of indicator	Proportion of households using coal, wood, dung or kerosene as the main source of heating and cooking fuel.
Underlying definitions and concepts	This indicator is based on the assumption that use of kerosene, wood, coal or dung for heating and cooking tends to increase levels of exposure to indoor air pollution.
	Underlying definitions are:
	"Household": a single dwelling unit (e.g. a house or apartment) intended for permanent residence.
	"Use of coal, wood, dung or kerosene as the main source of heating and cooking fuel": the reliance on coal (or lignite), wood, dung or kerosene as the primary cooking and heating fuel in the home.
Specification of data needed	Number of households using coal, wood, dung or kerosene as the main source of heating and cooking fuel.
	Total number of households.
Data sources, availability and quality	Data on the number of households using coal, wood, dung or kerosene as the main source of cooking and heating fuel may be available from census statistics or household surveys, and in these cases are liable to be broadly reliable. In many cases, however, data will need to be collected as part of special surveys.
	Data on the total number of households should be available through national census statistics, although care is needed in relation to the definition of a "household" (e.g. how collective dwellings are classified).

Continued

Table I   Continued

Sources of indoor air pollution   An exposure indicator

Computation	The indicator can be computed as:
	$(H_c / H_t) * 100$
	where H_c is the number of households using coal, wood, dung or kerosene as the main source of cooking/heating fuel, and H_t is the total number of households.
	The indicator should normally be calculated for a specified census date.
Units of measurement	Percentage.
Interpretation	This indicator provides a general measure of differences or trends in exposure to air pollutants from indoor heating and cooking sources; a reduction in the percentage of homes relying on coal, wood, dung or kerosene may be taken to imply a reduced level of exposure.
	In applying and interpreting the indicator, however, it should be noted that:
	• it takes no account of use of other sources of indoor pollution (e.g. smoking, furnishings and solvents);
	• the indicator takes no account of the many other factors (e.g. lifestyle and ventilation behaviour) likely to affect exposures;
	• relationships with health outcome may be heavily confounded by other factors, including exposures to outdoor and occupational pollution, housing conditions and socio-economic factors.

Table J

Childhood morbidity due to acute respiratory illness	An effect indicator

Indicator profile

Issue	Air pollution
Rationale and role	The incidence of acute respiratory illness in young children has shown a marked increase in recent decades, in almost all countries of the world. Many possible risk factors have been identified which might account for this trend; one of the most important is exposure to air pollution both in the home and outdoors.

This indicator is intended to provide a measure of the effect of these exposures to air pollution in children. As such, it can be used:

- to monitor trends in acute respiratory illness in children in order to help prioritise policy action;
- to map the distribution of the disease in order to identify areas in need of special action;
- to help identify specific at-risk groups in order to target intervention;
- to analyse relationships between air pollution (and other risk factors) and respiratory health;
- to assess the effectiveness of intervention strategies (such as air pollution control, traffic management and awareness raising campaigns).

Linkage with other indicators	This indicator represents one in a chain of indicators that together describe the effects of air pollution on health:

- State: *Ambient concentrations of air pollutants in urban areas.*
- Exposure: *Sources of indoor air pollution.*
- Effect: **Childhood morbidity due to acute respiratory illness**; *Childhood mortality due to acute respiratory illness.*
- Action: *Capability for air quality management; Availability of lead-free petrol.*

Alternative methods and definitions	This indicator can be defined as the incidence of morbidity due to acute respiratory illness in children under five years of age. Because acute respiratory illness tends to be more common in boys than in girls, it can usefully be standardised by gender. Where the aim is to investigate relationships with potential causative factors, stratification on the basis of other variables (e.g. ethnicity) may also be appropriate.

Continued

Table J   Continued

Childhood morbidity due to acute respiratory illness   An effect indicator

	Variations on this indicator are possible, depending on the availability of morbidity data. Sales of respiratory medication (e.g. inhalers) can be used as a proxy, although this is non-specific to this age group. Registrations at asthma clinics may also provide a proxy. The indicator could also be compiled and presented for other, more specific categories of acute respiratory infection, for example: • acute lower respiratory infection (ALRI), i.e. an acute infection of the larynx, trachea, bronchi, bronchioles or lung; • acute upper respiratory infection (AURI), i.e. an acute infection of the nose, pharynx (throat) or middle ear. Similar indicators might also be developed for other age groups considered to be at risk (e.g. the elderly).
Related indicator sets	WHO Catalogue of Health Indicators: • Care-seeking for children with acute respiratory infections.
Sources of further information	WHO 1992 *The Measurement of Overall and Cause-specific Mortality in Infants and Children.* Report of a Joint WHO/UNICEF Consultation, 15–17 December 1992. World Health Organization, Geneva. WHO 1994 *Ninth General Programme of Work Covering the Period 1996–2001.* World Health Organization, Geneva. WHO 1994 *The Management of Acute Respiratory Infections in Children: Practical Guidelines for Outpatient Care.* Division for the Control of Diarrhoea and Acute Respiratory Disease, World Health Organization, Geneva. WHO 1996 *Catalogue of Health Indicators: A Selection of Health Indicators Recommended by WHO Programmes.* World Health Organization, Geneva. WHO 1997 *Health and Environment in Sustainable Development: Five Years After the Earth Summit.* World Health Organization, Geneva.
Involved agencies	WHO — Department of Child and Adolescent Health and Development (CAH) UNICEF

Example indicator

Definition of indicator	Incidence of morbidity due to acute respiratory infections in children under five years of age.

Continued

Table J   Continued

Childhood morbidity due to acute respiratory illness	An effect indicator

Underlying definitions and concepts	This indicator is based on the following definitions:
	"Acute respiratory infection (ARI)": an acute infection of the ear, nose, throat, epiglottis, larynx, trachea, bronchi, bronchioles or lung.
	"Total population of children under five years of age": number of live children less than five years of age at the mid-point of the year (or other survey period).
Specification of data needed	Number of cases of acute respiratory infection in children under five years of age.
	Total number of children under five years of age.
Data sources, availability and quality	Data on the number of cases of acute respiratory infection amongst young children may be obtainable from a number of different sources, including hospital admissions, GP records and special surveys. None of these sources is comprehensive and wholly free of bias. Furthermore, GP data are generally difficult to acquire. For most purposes, therefore, the best available data are likely to come either from hospital admissions records or from specially designed surveys. The former includes only the more severe cases, and will omit cases which are not referred to hospital (e.g. which are treated at home or by the GP). Special surveys are inevitably based on relatively small samples, and may also suffer from bias or inconsistency in reporting.
	Data on the total number of children under five years of age are available from national census statistics, and should be reliable, especially for census years. Inter-censal estimates may be made using vital registration data or demographic models, but may contain some uncertainties due to effects of migration. These are likely to be significant only at the small area scale.
Computation	The indicator can be computed as:
	$1{,}000 * (R_c / P_c)$
	where R_c is the total number of cases of acute respiratory infection in children under five years of age in the survey period (e.g. the last calendar year), and P_c is the total number of children under five years of age at the mid-point of that survey period.
Units of measurement	Number per thousand children under five years of age.

Continued

Table J   Continued

Childhood morbidity due to acute respiratory illness   An effect indicator

Interpretation	This indicator is intended to provide a measure of changes or differences in the incidence of acute respiratory infections as a result of exposure to air pollution. In this context, an increase in the morbidity rate may be taken to infer an increase in exposures; a reduction in morbidity may imply a decrease in levels or frequency of exposure.
	In practice, however, such interpretations are problematic. Exposure to air pollution is only one of many possible causes of acute respiratory infection; other risk factors include exposures to house dust mite, damp and mould in the home, food additives and pollen. Factors such as family history, sibling order and genetic predisposition are also important. Associations between the incidence of acute respiratory infection and air pollution are thus complex and highly confounded. Data on morbidity are also limited and often inconsistent, making comparisons between different countries or interpretations of trends potentially difficult. Many cases go unreported. Differences in the structure of the health service (e.g. the extent of provision of asthma clinics) and in diagnosis also affect the reported rates. Attempts to combine statistics from different sources pose difficulties because of differences in classification and possible double-counting of individual cases. As with all morbidity measures, therefore, this indicator needs to be interpreted with care.

Table K

Childhood mortality due to acute respiratory illness	An effect indicator

Indicator profile

Issue	Air pollution
Rationale and role	Acute respiratory illness is the single largest cause of mortality in children under five years of age. This indicator measures the health effect of acute respiratory mortality in the high risk group of under five year olds. As an indicator for environmental health it provides an indication of potential health effects associated with the important issues of air pollution (especially indoor and vehicle pollution) and other environmental issues such as crowding and socio-economic status. Death due to acute respiratory illness is most commonly associated with infection or obstruction of the lower respiratory tract (i.e. the larynx, trachea, bronchi, bronchioles or lung). By providing a measurement of mortality in the sensitive group of under five year olds, this indicator also provides an indirect indication of potential health effects in older age groups.

As a measurement of cause-specific mortality, this indicator can serve several purposes:

- to establish the relative public health importance of acute respiratory illness as a cause of death;
- to monitor trends over time and provide an early warning of the need for intervention;
- to map variations in acute respiratory illness as a basis for identifying areas requiring special interventions;
- to monitor the effectiveness of policies and other interventions aimed at reducing acute respiratory mortality;
- to help investigate associations between air pollution or other risk factors and mortality due to acute respiratory illness;
- to provide an indication of the potential for other diseases associated with the same environmental health issues. (An important example in developing countries is diseases such as chronic respiratory disease in women as a result of exposure to domestic indoor air pollution from coal and biomass burning.)

Linkage with other indicators	This indicator represents one in a chain of indicators that together describe the effects of air pollution on health:

- State: *Ambient concentrations of air pollutants in urban areas.*
- Exposure: *Sources of indoor air pollution.*

Continued

Table K   Continued

Childhood mortality due to acute respiratory illness   An effect indicator

---

- Effect: *Childhood morbidity due to acute respiratory illness;* **Childhood mortality due to acute respiratory illness.**
- Action: *Capability for air quality management; Availability of lead-free petrol.*

Alternative methods and definitions	This indicator can be defined as the annual mortality rate due to acute respiratory illness in children under five years of age. Because acute respiratory infections tend to be more common in boys than in girls, it can usefully be standardised by gender. Where the aim is to investigate relationships with potential causative factors, stratification on the basis of other variables (e.g. ethnicity) may also be appropriate.

The indicator could also be compiled and presented for other, more specific categories of acute respiratory illness, for example:

- acute lower respiratory infection (ALRI), i.e. an acute infection of the larynx, trachea, bronchi, bronchioles or lung;
- acute upper respiratory infection (AURI), i.e. an acute infection of the nose, pharynx (throat) or middle ear.

In this way, the indicator could be applied to monitor or investigate disease-specific mortality. In developing countries, this might focus on the problem of pneumonia associated with biomass/coal-burning and indoor air pollution. (Typically this will comprise a high proportion of deaths due to acute respiratory illness in these countries.) In developed countries the growing problem of asthma associated with vehicle air pollution may prompt use of asthma-specific indicators.

Similar indicators might also be developed for other age groups considered to be at risk (e.g. the elderly).

Related indicator sets	WHO Catalogue of Health Indicators: - Under-five deaths due to acute respiratory infections.
Sources of further information	WHO 1992 *The Measurement of Overall and Cause-specific Mortality in Infants and Children.* Report of a Joint WHO/UNICEF Consultation, 15–17 December 1992. World Health Organization, Geneva. WHO 1994 *Ninth General Programme of Work Covering the Period 1996–2001.* World Health Organization, Geneva. WHO 1994 *The Management of Acute Respiratory Infections in Children: Practical Guidelines for Outpatient Care.* Division for the Control of Diarrhoea and Acute Respiratory Diseases, World Health Organization, Geneva.

Continued

Table K   Continued

Childhood mortality due to acute respiratory illness   An effect indicator

---

	WHO 1996 *Catalogue of Health Indicators: A Selection of Health Indicators Recommended by WHO Programmes.* World Health Organization, Geneva.
	WHO 1997 *Health and Environment in Sustainable Development: Five Years After the Earth Summit.* World Health Organization, Geneva.
Involved agencies	WHO — Department of Child and Adolescent Health and Development (CAH)
	UNICEF

Example indicator

Definition of indicator	Annual mortality rate due to acute respiratory infections in children under five years of age.
Underlying definitions and concepts	This indicator is based on the following definitions:
	"Acute respiratory infection (ARI)": an acute infection of the ear, nose, throat, epiglottis, larynx, trachea, bronchi, bronchioles or lung.
	"Total population of children under five years of age": number of live children less than five years of age at the mid-point of the year (or other survey period).
Specification of data needed	Annual number of deaths of children under five years of age due to acute respiratory infections (ARI).
	Total number of children aged under five years at the mid-point in the survey year.
Data sources, availability and quality	Data on childhood deaths due to ARI, especially in developing countries, are rare. In some countries, data may be available from demographic surveillance systems or from household surveys and, in some cases, from vital registration or sample registration systems. In a number of countries, the demographic surveillance surveys have included a verbal autopsy module aimed at collecting information on the cause of death in children.
Computation	This indicator can be computed as:
	$1{,}000 * (M_c / P_c)$
	where M_c is the number of deaths due to ARI in children under five years of age, and P_c is the total number of children under five years of age.

Continued

Table K   Continued

Childhood mortality due to acute respiratory illness	An effect indicator

Units of measurement	Number of deaths per thousand children below the age of five each year.
Interpretation	This indicator may be interpreted to show trends or patterns in mortality due to ARI as a result of exposure to air pollution. An increase in mortality rates might imply higher exposures and worsening air pollution conditions; a reduction in mortality might imply a decrease in exposures and an improvement in air quality.
	For many reasons, however, such interpretations need to be made with care. Crucially, the association between ARI mortality and air pollution is not simple. Many other factors may cause ARI, including exposures to dust mite and other allergens in the home; factors such as family history of atopy and sibling order are also important. In developing countries, HIV and malaria are extremely important factors in either causing lower respiratory infection, or presenting as LRI. These may thus have a substantial effect on observed death rates. Mortality is also highly dependent upon the effectiveness of the health care system and availability of treatment. Indeed, in many developed countries, mortality rates for ARI have remained broadly stable over recent decades, despite a large increase in morbidity.

Table L

Capability for air quality management	An action indicator

Indicator profile

Issue	Air pollution
Rationale and role	Many of the risks to human health from air pollution can be addressed and resolved through air quality management. Strategies for air quality management may vary substantially, depending on the specific sources and types of pollution involved and the social, political and environmental context. In general terms, however, management is aimed at controlling emissions at source in order to reduce pollution levels and prevent pollution episodes. Important elements of an air quality management strategy may thus include: air quality standards (for both short- and long-term concentrations); monitoring systems; emission limits and controls; and specific land use, transport, energy and industrial policies aimed at reducing air pollution.
	This indicator is thus an action indicator, designed to assess the capability to implement policies and strategies for air quality management. Its main purposes are thus:
	• to allow comparisons between areas or countries in terms of their air quality management capability (e.g. to help identify and disseminate good practice or to identify areas where improvements are needed);
	• to monitor and assess the implementation of air quality management strategies.
Linkage with other indicators	This indicator represents one in a chain of indicators that together describe the effects of air pollution on health:
	• State: *Ambient concentrations of air pollutants in urban areas.*
	• Exposure: *Sources of indoor air pollution.*
	• Effect: *Childhood morbidity due to acute respiratory illness; Childhood mortality due to acute respiratory illness.*
	• Action: **Capability for air quality management**; *Availability of lead-free petrol.*
Alternative methods and definitions	Developing indicators that adequately assess management capability is invariably difficult. In this case, a valuable and widely applicable approach has been developed by MARC (1996). This is a compound indicator, incorporating scores for four separate components of management capability, assessed in terms of 14 sets of variables:

Continued

Table L   Continued

Capability for air quality management   An action indicator

- air quality measurement capacity (measured in terms of the capacity to measure chronic health effects, acute health effects, trends in pollutant concentrations, spatial distribution of pollutants, kerbside concentrations and data quality);
- data assessment and availability (measured in terms of the capacity to analyse and disseminate data);
- emissions estimates (measured in terms of source emissions estimates, pollutant emissions estimates, accuracy of the emissions estimates and availability of the emissions estimates);
- air quality management capability tools (measured in terms of the capacity to assess air quality acceptability and to use air quality information).

This approach is comprehensive and provides a good, encompassing measure of the capability for air quality management at the city or local level. It may, however, need to be customised to specific circumstances, for example, according to the geographic scale and administrative context, where the focus of attention is on ambient concentrations rather than emissions, or where interest focuses on one specific source of pollution (e.g. transport). It is also possible to calculate and report the different components or variables (or combinations of them) separately, if appropriate.

Related indicator sets	GEMS/AIR:
	• Management Capabilities Assessment Index.
Sources of further information	MARC 1996 *Air Quality Management and Assessment Capabilities in 20 Major Cities.* GEMS/AIR. Monitoring and Assessment Research Centre, London.
	WHO 1998 *Healthy Cities Air Management Information System (AMIS).* AMIS v. 2.0. CDROM. World Health Organization, Geneva.
Involved agencies	National air quality monitoring agencies
	National environment ministries
	UNEP
	WHO

Example indicator

Definition of indicator	Capability to implement air quality management.

Continued

Table L   Continued

Capability for air quality management	An action indicator

Underlying definitions and concepts	This indicator is a based on the Management Capabilities Assessment Index, developed by MARC (1996) on behalf of UNEP and WHO. In this example, the original index has been simplified and adapted by selecting and redefining a smaller subset of variables, which might be considered most relevant to its application at the regional or national level. The scores have also been modified from the original index (they thus total to 60 rather than 100). Key definitions are:
	"Air quality management capability": the existence, implementation and enforcement of instruments and measures aimed at controlling or reducing air pollution in the ambient environment.
	"Air quality standards": legally specified limits for specific air pollutants which should not be exceeded over the specified averaging time.
	"Emissions controls": legally specified limits for emissions from specific sources which should not be exceeded under the specified operating conditions.
Specification of data needed	Evidence of the following capabilities is required in order to support this indicator: • capability for monitoring and reporting on air quality; • capability for measurement/estimation and reporting of emissions; • existence of, and capability to enforce, air quality standards; • existence and enforcement of emission controls; • integration of air quality issues into planning procedures.
Data sources, availability and quality	Information on the existence of these instruments and measures.

Computation	The index is computed as:
	
	$\Sigma\,(C_i)$
	
	where C_i is the score for component i.
	
	The full list of components (i) are as follows.
	
	1. A network of continuous monitoring sites covering residential areas for the following pollutants:
	   • NO_2
	   • SO_2

Continued

Table L   Continued

Capability for air quality management	An action indicator

- PM
- CO
- lead
- O_3.

Score 1 for each pollutant [max = 6].

2. Open access to air quality information through:
   - annual published reports
   - newspapers
   - the Internet.

   Score 1 for each pollutant [max = 3].

3. Publication of air quality warnings during pollution episodes.

   Score 3 if present [max = 3].

4. Requirement to measure and report emissions from:
   - major combustion sources
   - large industrial sources
   - other point emission sources.

   Score 1 for each source [max = 3].

5. Detailed emission inventories for NO_2, SO_2, PM, CO, metals (e.g. lead) and VOCs covering emissions from:
   - industrial sources
   - transport sources
   - domestic sources
   - other sources.

   Score 0.25 for each source and 1 for each pollutant;
   score calculated as sum of (source score x pollutant score)
   [max = 6].

6. Short-term (e.g. maximum daily) standards for:
   - NO_2
   - SO_2
   - PM
   - O_3
   - CO.

   Score 1 for each pollutant [max = 5].

7. Long-term (e.g. mean annual) standards for:
   - NO_2
   - SO_2
   - PM
   - lead.

   Score 1 for each pollutant [max = 4].

Continued

Table L   Continued

Capability for air quality management	An action indicator

8. Regulations to enforce compliance with air quality standards.
Score 3 if present [max = 3].

9. Arrangements to review and update air quality standards on a regular basis.
Score 3 if present [max = 3].

10. Emission controls for:
   • new road vehicles
   • domestic dwellings
   • industrial premises.
   Score 2 for each source [max = 6].

11. Availability of unleaded petrol
Score 3 if present [max = 3].

12. Requirement for testing of road vehicles, including testing of emissions, at a frequency of at least every five years for:
   • public service vehicles
   • heavy goods vehicles
   • cars.
   Score 2 for each group [max = 6].

13. Formal requirements for local air quality management strategies.
Score 3 if present [max = 3].

14. Requirements for air quality issues to be addressed as part of:
   • industrial development
   • major road developments.
   Score 3 for each type of development [max = 6].

Units of measurement   Ordinal score (0 – 60).

Interpretation   This indicator provides a general measure of the capability for air quality management; an increase in the score may thus be taken as a broad indication of increased capability, a reduction the reverse. Like all compound indicators, however, this one needs to be interpreted with care, for the final score is the sum of many different components. Areas with the same indicator score, therefore, do not necessarily have the same capability profile for air quality management. It is consequently important to examine the components of the indicator in drawing conclusions from the measure.

Table M

Availability of lead-free petrol	An action indicator

Indicator profile

Issue	Air pollution
Rationale and role	Vehicle fuel represents a major source of exposure to lead, traditionally accounting for 80–90% of the total lead concentration in the atmosphere. Other important sources are lead smelting, battery manufacture and refuse incineration. Chronic exposure to lead in the atmosphere is known to have a wide range of health effects, including raised blood pressure, disorders of the nervous system and haematological effects. In children, exposures are known to be associated with behavioural and learning difficulties.

The provision of unleaded petrol is one of the most effective and widely used methods of reducing lead emissions and thereby reducing human exposures. This indicator provides a measure of the action, and as such can be used to:

- monitor progress towards policy targets and goals on reducing lead in petrol;
- compare regions or countries in terms of their policies on lead pollution and exposure reduction;
- identify potential at-risk populations because of their raised exposure to lead in the atmosphere;
- analyse the effects of reductions in the use of leaded fuel in the environment and human health.

This indicator can also be used as a proxy measure of potential exposure to lead, especially in broad-scale studies.

Linkage with other indicators	This indicator represents one in a chain of indicators which together describe the effects of air pollution on health:

- State: *Ambient concentrations of air pollutants in urban areas.*
- Exposure: *Sources of indoor air pollution.*
- Effect: *Childhood morbidity due to acute respiratory illness; Childhood mortality due to acute respiratory illness.*
- Action: *Capability for air quality management;* **Availability of lead-free petrol.**

Alternative methods and definitions	This indicator can be defined as the percentage (by volume) of total petrol consumption provided by unleaded fuel. Alternative versions of the indicator might also be computed, for example, by using either population or surface area as the denominator (e.g. unleaded petrol consumption per head of population or

Continued

Table M   Continued

Availability of lead-free petrol	An action indicator

<table>
<tr><td></td><td>per km²). This would have the advantage of allowing for differences in the total volume of fuel consumed. Use of population as a denominator provides an indicator of the rate of consumption, and thus tends to highlight regions with high per capita usage of unleaded petrol. Use of area as the denominator provides an indicator of the intensity of consumption, and thus tends to highlight regions which potentially have high levels of emissions and higher atmospheric concentrations of lead.</td></tr>
</table>

per km^2). This would have the advantage of allowing for differences in the total volume of fuel consumed. Use of population as a denominator provides an indicator of the rate of consumption, and thus tends to highlight regions with high per capita usage of unleaded petrol. Use of area as the denominator provides an indicator of the intensity of consumption, and thus tends to highlight regions which potentially have high levels of emissions and higher atmospheric concentrations of lead.

Related indicator sets
GEMS/AIR Management Capabilities Assessment Index:
- Unleaded petrol available in the city.

Sources of further information
MARC 1996 *Air Quality Management and Assessment Capabilities in 20 Major Cities*. GEMS/AIR. Monitoring and Assessment Research Centre, London.

WHO 1992 *Human Exposure to Lead*. Report on the Human Exposure Assessment Locations (HEAL) Programme Meeting held in Bangkok, Thailand, 16–19 November 1992. World Health Organization, Geneva.

WHO 1994 *Ninth General Programme of Work Covering the Period 1996–2001*. World Health Organization, Geneva.

WHO 1995 *Inorganic Lead*. Environmental Health Criteria Series, Number 165. Published under the joint sponsorship of the United Nations Environment Programme, the International Labour Organisation, and the World Health Organization. World Health Organization, Geneva.

http://www.who.int/dsa/cat97/zehc1.htm

Involved agencies
WHO
Petroleum companies

Example indicator

Definition of indicator
Consumption of lead-free petrol as a percentage of total petrol consumption.

Underlying definitions and concepts
This indicator is based on the assumption that leaded fuel represents one of the main sources of exposure to lead in the atmosphere, and thus a significant health risk. Underlying definitions are:

"Unleaded petrol consumption": total sales (volume) of petrol not containing lead.

"Total petrol consumption": total sales of all petrol (by volume).

Continued

Table M   Continued

Availability of lead-free petrol	An action indicator

Specification of data needed	Volume of unleaded petrol sold. Total volume of petrol sold.
Data sources, availability and quality companies.	Data on the amounts of petrol sold are usually available from national statistics, and are typically derived either from trade data, taxation data or the sales data of the petroleum These data are reasonably reliable at the national level; at the regional/local level, however, they may be difficult to acquire (for reasons of commercial confidentiality) and may be less accurate.
Computation	The indicator can be computed as: $(G_u / G_t) * 100$ where G_u is the total volume of unleaded petrol sold, and G_t is the total volume of all petrol sold.
Units of measurement	Percentage.
Interpretation	This indicator is relatively simple to interpret, in that sales of unleaded petrol are influenced largely by policy action. In particular, differential taxation of fuels on the basis of their lead content is effective in controlling consumption. Nevertheless other factors affect consumption of unleaded fuels, including vehicle design and performance (both of which may be determined by manufacturers beyond the area of interest). Therefore, changes in sales of unleaded fuels should not necessarily be seen as evidence of the direct effects of policy action. When used as an indicator of exposure, it is also important to recognise that many other sources of exposure may occur, including industrial activity and coal combustion, both of which might be important locally. Recycling of lead in dust also means that relatively long delays may occur between reductions in use of leaded fuels and changes in atmospheric concentrations or human exposures.

Index